Bioproducts Processing
Technologies for the Tropics

Edited by M.A. Hashim

Institution of Chemical Engineers, Rugby, UK

Bioproducts Processing
Technologies for the Tropics

Orders for this publication should be directed as follows:

Institution of Chemical Engineers,
Davis Building,
165–189 Railway Terrace, RUGBY,
Warwickshire CV21 3HQ, UK

Tel: +44 1788–578214
Fax: +44 1788–560833

Bioproducts Processing
Technologies for the Tropics

A four day symposium organised by the Institute of Advanced Studies (University of Malaya) with the co-operation of the Institution of Chemical Engineers and the Institute of Engineers Malaysia and held at the Institute of Advanced Studies, Kuala Lumpur, Malaysia 4 –7 January 1994.

INSTITUTION OF CHEMICAL ENGINEERS

SYMPOSIUM SERIES No. 137
ISBN 0 85295 330 5

Printed by The Chameleon Press Limited, 5–25 Burr Road, London SW18 4SG, UK

Preface

This volume in the IChemE Symposium Series is a collection of selected papers presented at the International Symposium on Bioproducts Processing in January 1994. It spans the processing of biological materials upstream to downstream. The symposium itself represented the successful efforts of the Institute of Advanced Studies (University of Malaya), the Institution of Chemical Engineers and the Institution of Engineers Malaysia in bringing together, for the first time in Malaysia, over 180 participants from 18 countries.

All those who worked to make the symposium a success acknowledge the generous financial contributions of 25 agencies which include the Ministry of Science, Technology and the Environment (Government of Malaysia), Malaysia Airlines, UNESCO, UNEP, The British Council and the Government of Canada. I am indebted to the members of my Organising Committee for their untiring efforts. Finally, special thanks are due to the members of the Editorial Board for helping me get the manuscripts into their final form.

M.A. Hashim

Contents

Paper 1 Page 1 — Combined fermentation and radiometric studies to elucidate the mechanism of sucrose uptake by *Saccharomyces cerevisiae*
P.K. Mweisigye and J.P. Barford *(University of Sydney, Australia)*

Paper 2 Page 9 — An overview of process studies on bioconversion of oil palm waste into useful products
B.M.N. Mohd Azemi, A.A. Astimar, M. Anis and K. Das *(Universiti Sains Malaysia, Malaysia)*

Paper 3 Page 19 — Fermentative production of natural food colorants by the fungus *Monascus*
Y.K. Lee, D.C. Chen, B.L. Lim *(National University of Singapore, Singapore)*, H.S. Tay and J. Chua *(Food Biotechnology Centre, SISIR, Singapore)*

Paper 4 Page 25 — Microbial conversions of agro-waste materials to high-valued oils and fats
C. Ratledge *(University of Hull, United Kingdom)*

Paper 5 Page 35 — Performance evaluation and modifications of a Upflow Anaerobic Sludge Blanket (UASB) reactor treating sugar industry effluent: a case study
M.G. Grasius, L. Iyengar, C. Venkobachar *(Indian Institute of Technology-Manpur, India)* and H.B. Singh *(Hydro Air Tectonics, India)*

Paper 6 Page 43 — Growth and product formation of *Ankistrodesmus convolutus* in an air-lift fermenter
W.L. Chu, S.M. Phang and S.H. Goh *(University of Malaya, Malaysia)*

Paper 7 Page 51 — High productivity of thermostable xylanase free of cellulase: a promising system for large scale production
M.M. Hoq, C. Hempel and W.D. Deckwer *(Gesellschaft für Biotechnologische Forschung mbH, Germany)*

Paper 8 Page 59 — Modelling and optimisation of cell cultures
C. Sanderson, J.P. Barfod and G. Barton *(University of Sydney, Australia)*

Paper 9 Page 69	Protein enrichment of corn stover using *Neurospora sitophila* U.C. Banerjee, Y. Chisti and M. Moo-Young (University of Sydney, Australia)
Paper 10 Page 79	Adsorption kinetics of lyzosyme on the cation exchanger Fractogel TSK SP–650(M) M.A. Hashim, K.H. Chu and P.S. Tan *(University of Malaya, Malaysia)*
Paper 11 Page 87	The effect of impeller configuration on biological performance in non-Newtonian fermentations M.J. Kennedy and R.J. Davies *(New Zealand Institute of Industrial Research and Development, New Zealand)*
Paper 12 Page 101	Preparation and characterisation of activated carbon derived from palm oil shells using a fixed bed pyrolyser M. Normah *(Universiti Teknologi Malaysia, Malaysia)*, M.G.S. Yap and W.J. Ng *(National University of Singapore, Singapore)*
Paper 13 Page 111	Production of adriamycin and oxytetracycline by genetic engineered *Streptomyces* with palm oil and palm kernel oil as carbon sources C.C. Ho, I. Melor, L.M. Ong, A. Sarimah, M.P.O.K. Cheong, S.K. Lee, C.C. Yap and E.L. Tan *(University of Malaysia, Malaysia)*
Paper 14 Page 121	Production of recombinant proteins by high productivity mammalian cell fermentations P.P. Gray and K. Jirasripongpun *(University of New South Wales, Australia)*
Paper 15 Page 127	Effects of alkali, cellulase and cellobiase on the production of sugars from palm waste fibre M. Anis, K. Das and N. Ismail *(Universiti Sains Malaysia, Malaysia)*
Paper 16 Page 135	Separation techniques in industrial bioprocessing Y. Chisti and M. Moo-Young *(University of Waterloo, Canada)*
Paper 17 Page 147	Production of carbohydrate-based functional foods using enzyme and fermentation technologies M.J. Playne *(Dairy Research Laboratory, Australia)*

Paper 18 Page 157 — Gas holdup correlation for aerated stirred vessels
R. Parhasarathy, N. Ahmed and G.J. Jameson *(University of Newcastle, Australia)*

Paper 19 Page 169 — Engineering and microbiological aspects of the production of microbial polysaccharides: xanthan as a model
E. Galindo *(National University of Mexico, Mexico)*

Paper 20 Page 179 — A short study on the biodegradation of waste wax from a marine oil terminal
S.T. Nesaratnam *(The Open University, United Kingdom)*

Paper 21 Page 187 — Model prediction and verification of a two-stage high-rate anaerobic wastewater treatment system subjected to shock loads
M. Romli, J. Keller, P.L. Lee and P.F. Greenfield *(University of Queensland, Australia)*

Paper 22 Page 195 — The degradation of oil palm trunk using enzymes form *Trichoderma reesei* QM 9414
A. Putri Faridatul, A.K. Azizol and K.C. Khoo *(Forest Research Institute Malaysia, Malaysia)*

Paper 23 Page 203 — The process engineering of vegetable seeds; priming, drying and coating
A.W. Nienow, W. Bujalski *(University of Birmingham, United Kingdom)*, R.B. Maude and D. Gray *(Horticultural Research International, United Kingdom)*

Paper 24 Page 209 — Bioremediation of process waters contaminated with selenium
L. Riadi and J.P. Barford *(University of Sydney, Australia)*

Paper 25 Page 219 — An aqueous two-phase partitioning system of PEG-enzyme and tannic acid for recovery and reuse of cellulase in palm waste saccharification
K. Das, G. Dutta and M.A. Amiza *(Universiti Sains Malaysia, Malaysia)*

Paper 26 Page 227 — Computer control of fermentation processes by AI technologies
T. Yoshida *(Osaka University, Japan)*

Page 237 — Index

COMBINED FERMENTATION AND RADIOMETRIC STUDIES TO ELUCIDATE THE MECHANISM OF SUCROSE UPTAKE BY *SACCHAROMYCES CEREVISIAE.*

Patrick K . Mwesigye and John P . Barford
Department of Chemical Engineering, University of Sydney, NSW 2006, Australia.

Saccharomyces cerevisiae cells were grown on a mixture of sucrose and glucose and were found to utilise sucrose without first hydrolysing it. A batch study in which invertase was added to sucrose-limited media at mid-exponential phase of growth resulted in a decrease of specific growth rate from 0.50 h^{-1} to 0.44 h^{-1}. In radiometric studies, the uptake of labelled sucrose was higher in adapted cells (14 days) than in unadapted cells (2 days). These results indicate an adaptable sucrose transport system in yeast by which sucrose can be directly transported into the cells.

1. INTRODUCTION

The utilisation of multiple sugars by yeast has importance in a range of large scale biotechnology industries such as brewing, baker's yeast and alcohol fuel production. Fundamental understanding of the mechanism of such uptakes has been obtained by a range of individual experimental techniques (1, 5, 13, 14). It has been established that in the initial stages of fermentation, sucrose is hydrolysed to glucose and fructose by the action of the enzyme invertase prior to the sugars being transported across the cell membrane (7). Avigad (1) and Santos *et al.* (14) showed that yeast cells intracellularly accummulated [U-^{14}C] sucrose and studies with protoplasts have postulated hydrolysis outside the cell (7) and direct sucrose uptake (9). Orlowski and Barford (13) used the results of fermentative growth of *S.cerevisiae* on a sucrose-limited media to infer a direct uptake of sucrose into the yeast cell. A model simulating growth on sucrose substrate has been developed (4) and takes into account both the direct uptake of sucrose molecule into the yeast cells and the hydrolysis of sucrose to glucose and fructose prior to entry into the cell. The saccharides, maltose (15) and α-methylglucoside (11) have been reported to be actively transported into the cells.

Despite a large body of experimental work, no definitive understanding of mechanism of sucrose utilisation by *S. cerevisiae* exists. In this study, a range of different experimental methods has been used to give valuable insights into this mechanism. It has been found that, in addition to hydrolysis outside the cell by invertase, sucrose molecules can be transported directly into yeast cells.

2. MATERIALS AND METHODS

2.1 Yeast

The strain *Saccharomyces cerevisiae* 248 UNSW 703100 (2, 3) was used. This yeast was maintained on slopes containing : malt extract, 5 g/l ; glucose, 4 g/l; agar, 15 g/l and vegemite (a local commercial yeast extract, Kraft Foods Ltd., Melbourne, Australia), 4 g/l. The culture was stored at 4°C.

2.2 Minimal Medium (MMA)

Fully defined minimal medium with the following composition (based on 10 g/l sugar solution): $(NH_4)_2SO_4$, 5 g/l; KH_2PO_4, 3.4 g/l; Na_2HPO_4 , 0.41 g/l; citric acid, 0.21 g/l; $MgCl_2 . 6H_2O$, 0.014 g/l; $FeCl_3 . 6H_2O$, 7.0 x 10^{-3} g/l; $CaCl_2$, 3.5 x 10^{-3} g/l; $MnCl_2 . 4H_2O$, 2.5 x 10^{-3} g/l; $CoCl_2$, 0.6 x 10^{-3} g/l; H_3BO_3, 0.2 x 10^{-3} g/l; $CuCl_2 . 2H_2O$, 0.4 x 10^{-3} g/l; $ZnCl_2$, 2 x 10^{-5} g/l; biotin, 0.08 x 10^{-3} g/l ; p-aminobenzoic acid, 8 x 10^{-3} g/l; riboflavin-5-phosphate, 8 x 10^{-3} g/l; pyridoxine hydrochloride, 0.016 g/l; thiamine, 0.016 g/l; inositol, 0.08 g/l. The minimal media was prepared at 10 times the above concentrations (10 x MMA); filter- sterilised (0.2 μm, Millipore) and stored at 4°C. Fermentation media was prepared by adding 10 x MMA to sterilised sugar solution in the ratio 1: 9.

2.3 Inoculum Preparation

The yeast cells were transferred from agar slopes into a 250-ml flask containing about 100 ml of liquid sucrose-limited growth medium and adaptation carried out for either 2 or 14 days as previously described (5,12). An actively growing inoculum, taken during exponential phase, was used in all experiments.

2.4 Fermentation Conditions

All experiments were carried out under strictly defined conditions of pH at 5.0 ± 0.1, temperature at 30.0 ± 0.5°C and oxygen was maintained at greater than 75% saturation. These parameters were all controlled on-line using an FC-4 control system (Real Time Engineering, Sydney, Australia).

2.5 Sugar Analysis

At specified time intervals during fermentation, 20 ml of yeast cell suspension was withdrawn. The samples were filtered through 0.45 μm membrane filter papers (Gelman Sciences). The filter papers were dried in an oven at 105°C for 30 minutes. After cooling in a dessicator, the filter papers were weighed for dry weight determination. The filtrate was analysed for sugars using high performance liquid chromatography (Waters Shodex SC 1011 column). Detection of sugars was achieved by a differential refractometer (model R401). Data was collected, stored and integrated using Delta Junior software (Digital Solutions, Brisbane, Australia).

2.6 Calculation of Specific Growth Rate

Specific growth rates of the cells during fermentation were calculated from the slope of semilogarithmic plots of dry weight versus time by linear regression.

2.7 Invertase Solution

Invertase was dissolved in 0.32 M citrate buffer (pH 4.6) to a concentration of 5 mg/ml. Twenty millilitres of this solution was added to the fermenter at mid-exponential phase of growth so that all the sucrose still present was hydrolysed to glucose and fructose.

2.8 Measurement of Sucrose Uptake

Exponentially growing cells were harvested by centrifuging at 3500 x g for 15 minutes. The cells were washed twice with ice-cold MMA and suspended in the same media to a cell density of 50 - 70 mg [dry weight] of yeast cells ml^{-1} at room temperature. Uptake studies were initiated by adding 50 μl of this cell suspension to 50 μl of labelled sucrose solution (0.2 - 0.4 μCi). The reaction mixture, in a final volume of 100 μl, contained 2 - 4 mg [dry weight] cells, 2 mM each of unlabelled glucose and fructose, MMA (same final concentration as of growth media) and labelled sucrose of the desired concentration. After incubation for 10 seconds at room temperature, 10 ml of ice-cold MMA was added and filtered through 0.45 μm membrane filters (Millipore, 25 mm) and washed with 10 ml of the same ice-cold MMA. The filters were then transferred into aqueous counting scintillant (ACS II) and radioactivity determined on liquid scintillation counter (Packard Model 4000 Minimax Tri-Carb).

Controls were run in parallel to subtract non-specific binding of labelled sugar by boiling the cell suspension for 15 minutes in a water bath at 100°C. The boiled cells were then cooled on ice and labelled sugar added. Uptake assay was carried out as above. The control values of uptake have already been subtracted for data presented.

2.9 Chemicals

[U-^{14}C] sucrose (560 mCi/mmol), [U-^{14}C] glucose (287 mCi/mmol), [U-^{14}C] fructose (304 mCi/mmol) and liquid scintillant (ACS II) were obtained from Radiochemical Centre, Amersham, U.K.; sucrose, glucose and fructose (analytical grade) from Ajax Chemicals, Sydney, Australia and invertase from Boehringer Mannheim. All other chemicals were of analytical grade.

3. RESULTS AND DISCUSSION

3.1 Growth on Glucose and Sucrose Mixture

The yeast cells were grown on a mixture of sucrose (8 g/l) and glucose (12 g/l) in a batch fermenter after adapting the cells on sucrose for either 2 or 14 days. The results of sugar analysis are shown in Figure 1. After 2 days adaptation, the cells were found to utilise sucrose after almost all the glucose was used up (Figure 1a). The overall specific growth rate was 0.45 h^{-1} which was the same as that obtained by Orlowski (12) when growing the cells on glucose alone. After 14 days adaptation, sucrose and glucose were utilised simultaneously during exponential phase of growth although glucose was utilised at a faster rate (Figure 1b). The overall specific growth rate was 0.47 h^{-1} which was significantly less than that obtained when cells are grown on sucrose alone after 14 days adaptation (see Table 2). In both cases, there was no fructose detected in the medium which implies that sucrose was fermented without first being hydrolysed to glucose and fructose. This also means that the secretion of invertase was supressed by glucose even after the cells were adapted on sucrose for 14 days (fully adapted). Catabolite inhibition of invertase and regulation of its synthesis have been reported (8, 10) and they have also been demonstrated in this work. Most significant is the fact that sucrose was completely consumed by the yeast cells without first being hydrolysed to its constituent monosaccharides. This means that it is no longer a necessary condition that for yeast cells to grow on sucrose, it has first to be hydrolysed to glucose and fructose. This suggests that there is a sucrose transport system in yeast by which it can be transported directly into the cells.

3.2 Radiometric Studies

Since the initial uptake rates were found to be highest during the first 10 seconds of incubation and effects of hydrolysis and metabolism minimum, radiometric studies on sucrose transport were carried out by incubating the cells for only 10 seconds. Table 1 shows the uptake of sucrose by fully adapted (14 days) and unadapted (2 days) yeast cells. The uptake values are an average from four different experiments. The uptake by adapted cells was higher at all sucrose concentrations used. This means that the sucrose transport system is more pronounced after adapting the cells. This is in agreement with results obtained from growth of the yeast cells on sucrose in which the growth rate increased after adapting the cells for 14 days (12). Burger *et al.* (6) showed that the physiological state of the yeast cells is of importance for the rate of sugar penetration into the cells. This means that if the yeast cells are allowed enough time to adapt on sucrose, the uptake is increased and this should have considerable advantages in industries that utilise sucrose as one of the main substrates. It can, therefore, be concluded that the sucrose transport system is inducible and depends on the time allowed for adaptation, the conditions under which adaptation occurs and the physiological state of the yeast cells.

3.3 Studies involving Invertase Addition to Growing Cultures

The yeast cells were grown on a sucrose-limited medium before invertase was added at mid-exponential phase. The results of growth before and after invertase addition are shown in Figure 2a. As can be observed, there was a decrease in the growth rate after addition of invertase. The average specific

TABLE 1: Uptake of labelled sucrose by fully adapted and unadapted yeast cells. The cells were adapted on sucrose for either 2 or 14 days. Uptake studies were initiated by adding labelled sucrose to cells and assay carried out as described in Section 2.8. The values represent average from four trials.

[sucrose] (mM)	Sucrose uptake (nmol/mg dry wt)	
	2 days	14 days
1.0	0.203 ± 0.030	0.467 ± 0.104
2.0	0.434 ± 0.090	0.691 ± 0.060
5.0	0.952 ± 0.089	1.572 ± 0.169
8.0	1.475 ± 0.034	2.370 ± 0.056
10.0	1.847 ± 0.043	2.883 ± 0.307

TABLE 2: Specific growth rates of *S. cerevisiae* before and after invertase addition to sucrose-limited batch culture. Invertase solution was added to the batch culture at mid-exponential phase in all experiments except in experiments 6 and 7 which acted as controls. The cells were adapted on sucrose for 14 days before inoculation into fermenter.

Experiment	specific growth rate (h^{-1})	
	before addition	after addition
1	0.52	0.45
2	0.50	0.45
3	0.48	0.44
4	0.51	0.42
5	0.50	0.44
6	0.51	-
7	0.52	-

growth rate was 0.50 h^{-1} when the cells were growing on sucrose and decreased to 0.44 h^{-1} after addition of invertase (Table 2). Figure 2b illustrates the sugar(s) profile before and after invertase addition. This profile, which was characteristic of all the five experiments carried out, indicates that all the sucrose was hydrolysed to glucose and fructose after addition of invertase. Therefore, after addition of invertase, the cells were growing on a resulting mixture of glucose and fructose only.

A decrease in growth rate after all the sucrose has been hydrolysed to glucose and fructose is very significant. It indicates that the contribution of sucrose in achieving a higher growth rate has been eliminated by the hydrolysis process. This again supports the proposition of a direct uptake component for sucrose utilisation by yeast.

3.4 Conclusion

A combination of fermentation and radiometric methods has been used to provide definitive evidence for direct uptake of sucrose by growing cultures of *S.cerevisiae*. Taken in totality, these experiments are the first to conclusively demonstrate the direct uptake of sucrose by actively growing yeast cells.

Acknowledgement. The partial funding of this work by the Australian Research Council is appreciated. We thank Departments of Biochemistry and Microbiology, University of Sydney, for providing the liquid scintillation counter.

4. REFERENCES

1.Avigad, G., 1960, Biochim. Biophys. Acta , 40, 124-134.
2. Barford, J.P. and Hall, R.J.,1979, Biotechnol. Bioeng., 21, 609-626.
3. Barford, J.P. and Hall, R.J., 1979, J. Gen. Microbiol., 114, 267-275.
4. Barford, J.P., Phillips, J.P. and Orlowski, J.H., 1992, Bioprocess Eng., 7, 303-307.
5. Barford, J.P., Mwesigye, P.K., Phillips, J.P., Jayasuriya, D. and Blom, I., 1993, J. Gen. Appl. Microbiol., 39, 389-394.
6. Burger, M., Hejmova, L. and Kleinzeller, A., 1959, Biochem., 71, 233-242.
7. De la Fuente, G. and Sols, A., 1962, Biochim. Biophys. Acta, 56, 49-62 .
8. Elorza, M.V., Villanueva, J.R. and Sentandreu, R., 1977, Biochim. Biophys. Acta, 475, 103-112.
9. Friis, J. and Ottolenghi, P., 1959, Compt. Rend. Trav. Lab. Carlsberg, 31, 259-271.
10. Mormeneo, S. and Sentandreu, R., 1982, J. Bacteriol., 152, 14-18.
11. Okada, H. and Halvorson, H.O., 1964, Biochim. Biophys. Acta, 82, 538-548.
12. Orlowski, J.H., 1987, Ph.D thesis, University of Sydney, Australia.
13. Orlowski, J.H. and Barford, J.P., 1991, J.Gen. Appl. Microbiol., 37, 215-218.
14. Santos, E., Rodriguez, L., Elorza, M.V. and Sentandreu, R., 1982, Arch. Biochem. Biophys., 216, 652-660.
15. Serrano, R., 1977, Eur. J. Biochem. 80, 97-102.

Symbols

ACS	aqueous counting scintillant
MMA	minimal media

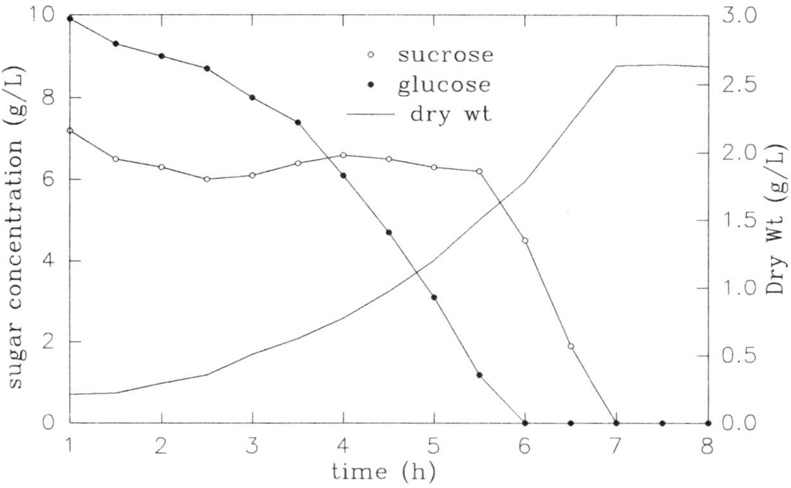

Fig. 1a Fermentation of sucrose and glucose mixture after 2 days adaptation on sucrose.

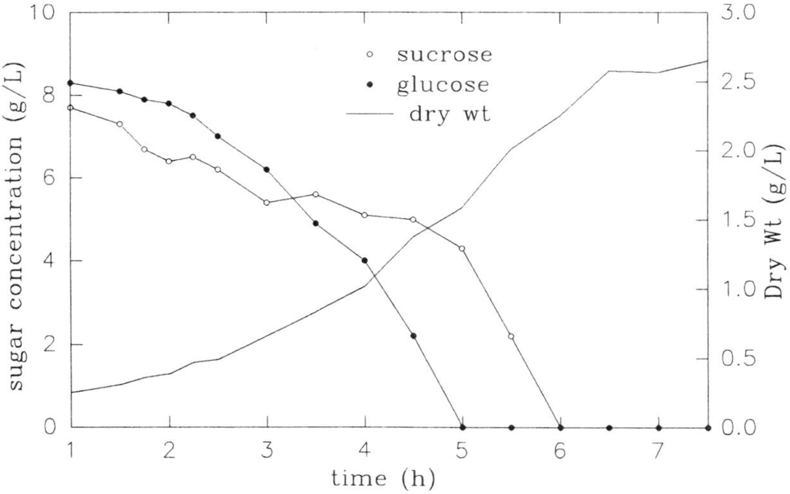

Fig. 1b Fermentation of sucrose and glucose mixture after 14 days adaptation on sucrose.

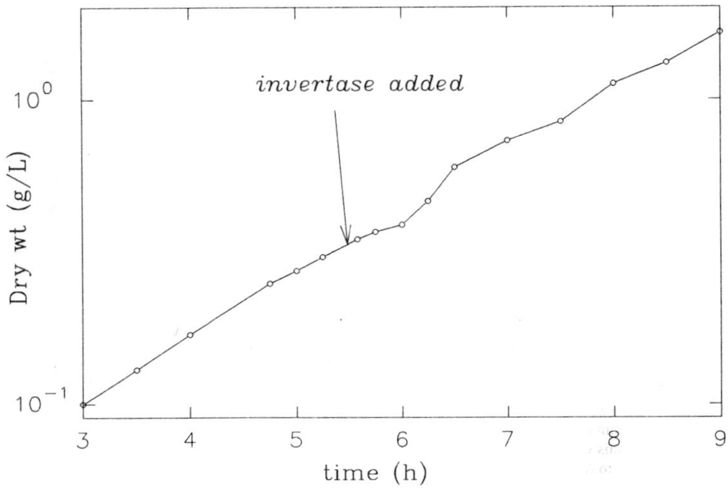

Fig. 2a Growth of *S.cerevisiae* on sucrose before and after invertase addition

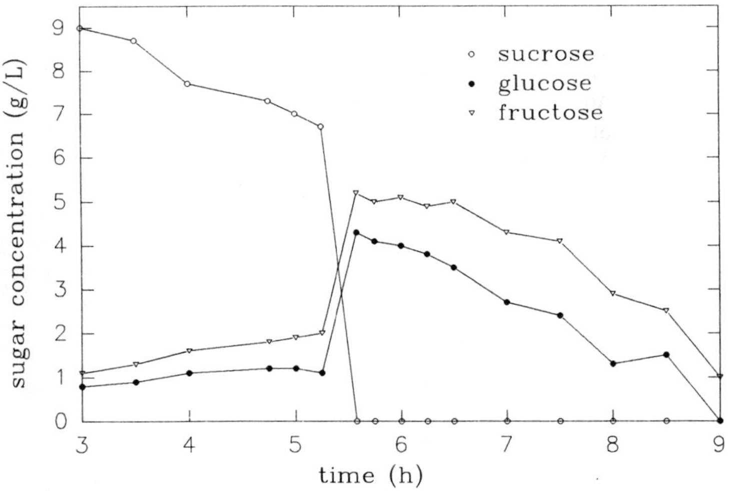

Fig. 2b Sucrose medium utilisation before and after invertase addition

AN OVERVIEW OF PROCESS STUDIES ON BIOCONVERSION OF OIL PALM WASTES INTO USEFUL PRODUCTS

Mohd. Azemi, B.M.N, Astimar, A. A, Anis, M and Das, K,* Division of Food Technology, School of Industrial Technology, Universiti Sains Malaysia, Minden, 11800 Penang, MALAYSIA
* Corresponding Author

Holocellulose, obtained from oil palm waste by soaking it in 1M NaOH for 2 h and then treating with chlorous acid at 70 - 75°C, produced high amount of xylose when it was prehydrolysed with 1M trifloroacetic acid and autoclaved at 121°C for 25 min. The total conversion of pentose to xylose was 95%. The xylose-rich prehydrolysate was found to serve as a good medium for *Candida tropicalis* ITM 3022 in producing high yield of xylitol. Fermentation under optimum conditions gave a xylitol production up to 20 g/l after 11 days of fermentation. Hydrolysis of the solid residue with cellulase and cellobiase enzymes gave a conversion as much as 90 % of the cellulose to glucose. The glucose-rich hydrolysate was then used as a medium for the production of glutamic acid with *Brevibacterium lactofermentum* ATCC 13869 yielding as much as 88 g/l (equivalent to 88% conversion based on the glucose used in the hydrolysate and 73% conversion based on the treated palm waste). A simultaneous study had also been done on the glucose-rich hyrolysate with *Kluveromyces fragilis* ITM 3039 under optimum conditions of pH 7.5, temperature of 30°C and agitation at 200 rpm. The concentration of this edible biomass reached its maximum after a fermentation period of 48 h to 5 g/l equivalent to 63% conversion of the potential glucose.

1. INTRODUCTION

In view of the vast quantities of oil palm cellulosic wastes available in Malaysia, a laboratory investigation into the potential use of the oil palm press fibre as a raw material for the production of single cell protein (SCP) and valuable chemicals, such as, xylitol and glutamic acid through fermentaion processes has been initiated. Normally, the empty bunches contain 30.5% dry matter, 2.5% oil and 63.0% water. After pressing, about 13.5% and 5.5% of the total weight of the fresh fruit bunches is converted into pressed fruit fibres and shells respectively. The dry matter of the pressed fruit fibres and shell is approximately 60% and 85% of their weight respectively as estimated by Mohd. Husin *et al* (1).

Solid oil palm wastes (palm bunch and fruit press fibres) contain approximately available hemicellulose (pentosan) 24%, cellulose 40%, lignin 21% and ash with extractives 15%. This delignified palm waste (holocellulose) after hydrolysis gives major fermentable sugars, such as, glucose and xylose through two-stage hydrolysis. These serve as feedstocks for a variety of fermentations producing value added products.

To hydrolyse cellulosic materials effectively with enzymes, pretreatment of these raw materials is neccesary to make the cellulose more accessible. The pretreatment is conducted under conditions that result in minimum degradation of the hemicellulose and cellulose and separates the lignin fraction in economical form. The yield of fermentable sugars is found to be markedly higher when the substrate is swelled by alkali treatment than that when it is simply boiled with alkali. Dilute NaOH treatment of lignocellulosic materials causes swelling, leading to an increase in internal surface area, decrease in the degree of polymerization, decrease in crystallinity, separation of structural linkages between lignin and carbohydrates and disruption of the lignin structure. Following this treatment, holocellulose can be prepared by solubilization and extraction of lignin through treatment with chloruos acid at 70 - 75 °C.

For selective hydrolysis of hemicellulose part, methods using 1M Trifluoroacetic Acid (TFA) at high temperature (121°C) has been developed by Astimar *et al* (2) and Low *et al* (7) which produce less or no toxic by-products that may impair cell growth. It was also found that this prehydrolysis could not significantly hydrolyse the cellulose part which was shown by low yield of glucose in the prehydrolysate. After prehdrolysis, which removes the pentose from the solid palm waste, the solid residue retained after separation of filtrate is mainly cellulosic residue. The rate of cellulose degradation is strongly affected by end-product inhibition caused by cellobiose and glucose, cellobiose being a stronger inhibitor in the hyrolysis process. To maximise glucose formation, it is necessary to use cellobiase enzyme alongwith the cellulase application.

Xylitol is a five carbon sugar alcohol which has an anticarciogenic property and does not cause acid formation as shown by Makinen (3) and is used as a sugar substitute for treatment of diabetes or when there is a deficiency of glucose-6-phosphate dehydrogenase indicated by Wang and Van Eys (4). Many yeast species and mycelial fungi possess the enzyme, xylose reductase, which catalyses the reduction of xylose to xylitol as the initial step in xylose metabolism as observed by Chiang and Knight (5). Preliminary studies done by Yap *et al* (6) show that *Candida tropicalis* (Institut Teknologi Mara, ITM 3022) grows well in the bagasse hemicellulose hydrolysate and gives a high production of SCP and xylitol. The ability of this strain to ferment full-strength hydrolysate was done by a continuous adaptation-selection technique which allows this strain to overcome the inhibitory effect of extraneous chemicals in the palm waste pentose hydrolysate. From the study of Low et al (7), it was observed that *C. tropicalis* produced a higher amount of xylitol which had overcome the inhibitory effect of the palm waste pentose hydrolysate.

Glutamic acid is one of the natural amino acids used in the food industry as a flavour enhancer. It plays a major role in the palatability and acceptibility of many foods. *Brevibacterium lactofermentum* ATCC 13869 is one of the important bacteria which produces glutamic acid. Studies were conducted to find out the optimum cultural conditions on the fermentation of palm waste cellulose hydrolysate and successful yield of glutamic acid was observed by Anis *et al* (8). Apart from this, a simultaneous study had also been done on the glucose-rich hydrolysate with *Kluveromyces fragilis* ITM 3039 under optimum conditions of pH 7.5, temperature of 30°C and agitation at 200 rpm for the production of single cell protein by Low *et al* (7).

2. MATERIALS AND METHODS

2.1 Raw Materials and Its Preparation

Oil palm press fibres used in this study were collected from Malpom Industry, Penang. After grinding the washed and dried raw material to proper size in the hammer mill (Retsch 20, Germany), the swelling was done by soaking it in 1M NaOH for 2 h. Holocellulose was prepared by solubilization and extraction of lignin through treatment with chlorous acid at 70-75°C. The solid residue was then filtered away and washed thoroughly with hot water until the alkali/acid was completely removed. The ratio between the substrate (treated palm waste) and the alkali was 1:20 (w/v).

2.2 Prehydrolysis of Solid Palm Waste and Fermentation to Xylitol

The delignified solid palm waste was then hydrolysed with 1M TFA at 10% (w/v) and autoclaved at various times (5 - 30 min) at 121°C. The filtrate (termed 'prehydrolysate') was then subjected to a vacuum distillation at 40°C for 3 h to remove TFA and to concentrate. The pH of the acid prehydrolysate (pH 0.3) was adjusted to 10 by adding CaOH. The mixture was then centrifuged to remove the precipitates and concentrated phosphoric acid was added to bring the pH to 4.0. After centrifugation, solid KOH was added to adjust the pH to 5 and precipitates were again removed by centrifugation. The treated prehydrolysate was then used as the fermentation medium. The strain used was an edible yeast, *C. tropicalis* ITM 3022. The fermentation was conducted at pH 5, temperature of 30°C for a period to reach maximum yields of products which were recovered after the period.

2.3 Saccharification and Fermentation of Cellulose Hydrolysate to Glutamic Acid

After prehydrolysis, the solid palm waste retained after separation of filtrate is mainly cellulosic residue. The solid concentration maintained for saccharification by enzyme was 1% (w/v) using 0.5 M acetate buffer at pH 4.8. Experiments were first conducted with cellulase enzyme (18 IU/ml) under controlled conditions of temperature and agitation at 48°C and 150 rpm respectively followed by reaction with cellobiase enzyme (5.4 IU/ml) under similar optimum conditions.

After 48 h of saccharification, this cellulose hydrolysate containing glucose was then used as the fermentation medium for *B. lactofermentum* which was found to be highly operative for better yield of the acid at pH 7.5, temperature of 30°C and agitation at 200 rpm. For optimisation of fermentation parameters, nutrient medium with following compositions were added (%, w/v): $MnSO_4.H_20$, 0.001; $FeSO_4.7H_20$, 0.001; K_2HPO_4, 0.2; $MgSO_4.7H_2O$, 0.2 and biotin 20 μg/l.

The outline of the sample preparation, pretreatment, acid prehydrolysis, enzymic hydrolysis and the fermentation processes are shown in Fig. 1.

2.4 Analytical Methods

A high performance liquid chromatography (HPLC, Waters Chromatography Div., Milford) using Sugar Pak column (6.5 mm x 300mm) was used for identification and quantification of monosaccharides and xylitol. Ethanol was analysed by Gas Chromatography (Shimadzu), total reducing sugars by DNS method after Miller (9), protein by Folin - Lowry method of Peterson (10) and single cell protein was calculated as dry cell weight after drying the centrifuged cell mass at 105°C until constant weight.

3. RESULTS AND DISCUSSION

3.1 Fermentation of Prehydrolysate to Xylitol and Edible Biomass

For saccharification of hemicellulose part, prehydrolysis was done using 1M TFA, autoclaved at 121°C for 25 minutes. The hemicellulose hydrolysate consisted principally of xylose (20 g/l) and some amounts of glucose (5 g/l) and arabinose (3 g/l). The total conversion of pentosan to xylose was 86%. On the other hand, the amount of xylose in hemicellulose hydrolysate was found to be higher from the sample when it was treated with 1M NaOH followed by bleaching with chlorous acid (Figure 2).

The production of xylose was equivalent to 95% conversion (g. xylose per 100 g. pentosans) or 23% conversion to xylose based on solid palm waste. Bleaching process together with preliminary treatment with soaking in 1M NaOH gave a final residue of white colour, termed as "holocellulose". A sole treatment with 1M NaOH or with the bleaching agent was found rather insufficient.

The ability of *C. tropicalis* to ferment full-strength hydrolysate was done by fermentation on the half-strength hydrolysate and transferring the inoculum into the full-strength hydrolysate. Direct fermentation of the hydrolysate with *C. tropicalis* produced maximum biomass up to 3.3 g/l (protein content of 1.2 g/l) together with ethanol and xylitol of 0.8 and 0.2 g/l respectively when fermented at optimum conditions of temperature of 30°C, pH 5.5, agitation at 200 rpm for 72 h (Figure 3). Thiamine hydrochloride (200 mg/ml) together with 5000 mg/ml of $MgSO_4.7H_2O$ were found to increase the biomass, ethanol and xylitol production as much as 54, 29 and 58 % respectively (Figure 4). Growth promoting factors, biotin at 0.5 μg/l and yeast extract at 1g/l had further increased respective yields of xylitol, biomass and ethanol up to 10, 9 and 8 g/l in 72 h fermentation. Period of fermentation was also important and when extended up to 11 days, xylitol and biomass yields were 20 and 11 g/l respectively minimising the ethanol to only 1.0 g/l. The mode of fermentation after final studies under optimal conditions of fermentations of the prehydrolysate by *C. tropicalis* is shown in Figure 5 .

3.2 Fermentation of Cellulose Hydrolysate to Glutamic Acid and Edible Cell Mass

The yield of glucose by saccharification of solid palm waste residue with cellulase (18 IU/ml) along with cellobiase (5.4 IU/ml) markedly increased the conversion up to about 90% within 48 h. With glucose concentration of the hydrolysate maintained at 10 g/l and with addition of nutrients, fermentation by *B. lactofermentum* under optimum conditions of pH 7.5, temperature of 30°C and agitation at 200 rpm, an yield of 88 g of glutamic acid (equivalent to 88% conversion based on the glucose used in the hydrolysate and 73% conversion based on treated palm waste) was achieved after 72 h which has surpassed earlier yield of 45% obtained by Rosma *et al* (11). The conversion was 25% higher than that of pure glucose (Figure 5).

A simultaneous study on the production of edible protein using *K. fragilis* had also been done by the fermentation of the glucose in that hydrolysate under optimum conditions of pH 7.5, temperature of 30°C and agitation at 200 rpm. Prior to the fermentation process, the glucose content of the hydrolysate was adjusted to ca. 10 g/l and the concentration of this edible biomass reached its maximum after a fermentation period of 48 h to 5 g/l equivalent to 63% conversion of the potential glucose (Figure 6).

ACKNOWLEDGEMENTS

The authors are grateful to 'Malayan Sugar Manufacturing (MSM) Co. Berhad' for funding this project.

REFERENCES

1. Mohamad, H, Zakaria, Z. Z and Hassan, A. H, 1985. Potentials of Oil Palm By- products as Raw Materials for Agro-based Industries. Procs., National Symposium on 'Oil Palm By-products for Agro-based Industries'. pp.7

2. Astimar, A. A, Das, K and Mohd Azemi, B. M. N, 1992. Improved Production of Edible Biomass by Fermentation of Resin Treated Xylose Hydrolysate of Palm Fibres. Pub. Procs, 15th Malaysian Microbiology Symposium on 'Microbial Diversity', Oct. 27 - 28, Penang, Malaysia, pp. 73

3. Makinen, K. K, 1978. Biochemical Principles of the Use of Xylitol in Medicine and Nutrition with Special Consideration of Dental Aspect. Exper. Suppl ., 30,7

4. Wang, Y. M and Van Eys, J. V, 1981. Ann. Res. Nutr., 30 , 1139

5. Chiang, C and Knight, S. G, 1960. Metabolism of D-xylose by Moulds. Nature 188, 79

6. Yap, B. T, Das, K and Astimar, A. A, 1991. Edible Biomass and Chemical Feed Stocks from Prehydrolysates of Cellulosic Food Waste. Procs., Fourth Chemical Congress of North America, Aug. 25 - 30, New York, USA

7. Low, E. C, Das, K and Astimar, A. A, 1993. Process Optimisation for Bioconversion of Palm Waste Pentosan into D-xylitol and Food Yeasts. Pub. Procs., Symp. on 'Asean- Australian Biotechnology Co-operation' under 11 th Australian Biotechnology Conf., Sept. 20 - 24, Perth, West Australia, pp. 165

8. Anis, M, Das, K and Ismail, N, 1993. Improved Yield of Glutamic Acid by Fermentation of Palm Waste. Pub. Procs., 16 th Malaysian Microbiology Symp on 'Advances in Microbial Research and Applications', Oct. 28 - 29, Pulau Langkawi, Malaysia, pp. 146

9. Miller, G. L, 1959. Use of Dinitrosalicylic Acid Reagent for Determination of Reducing Sugars, Anal. Chem., 31, 426

10. Peterson, G. L , 1979. Anal. Biochem., 100, 201-220

11. Rosma, A, Das, K, Anis, M and Ismail, N, 1993. Production of Glutamic Acid by *B. lactofermentum* from Palm waste Hydrolysate. Pub. Procs., 'Asean - Austr. Biotech. Co-op.' under 11 th Australian Biotechnology Conf., Sept. 20- 24, Perth, West Australia, pp. 222

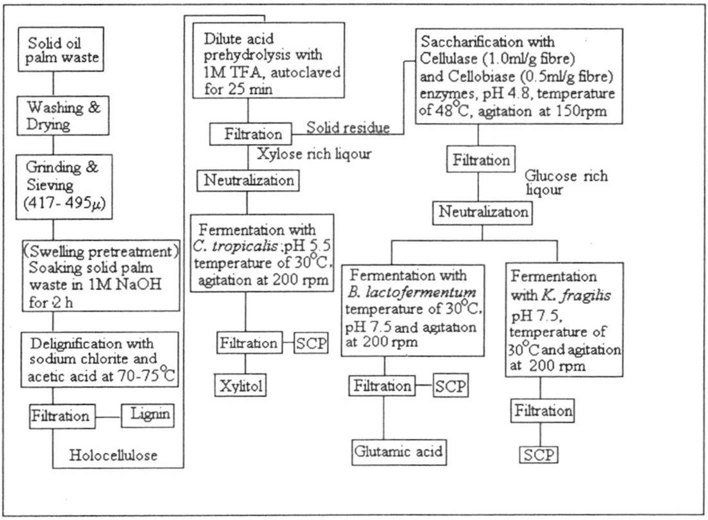

Fig. 1. A schematic diagram of a process of two-stage acidic and enzymic hydrolysis of solid oil palm waste and the fermentation of its hydrolysate for the production of SCP and other chemicals

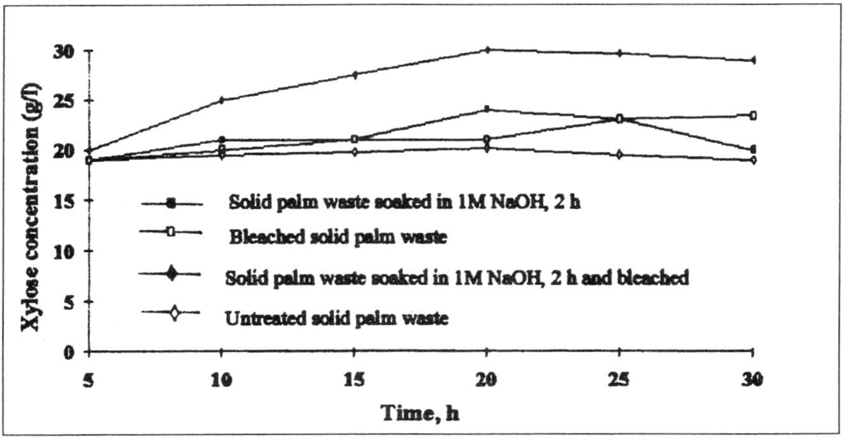

Fig. 2. Xylose yields from the acid hydrolysis of solid palm waste holocellulose with 1M TFA (autoclaved at 121 °C at various periods)

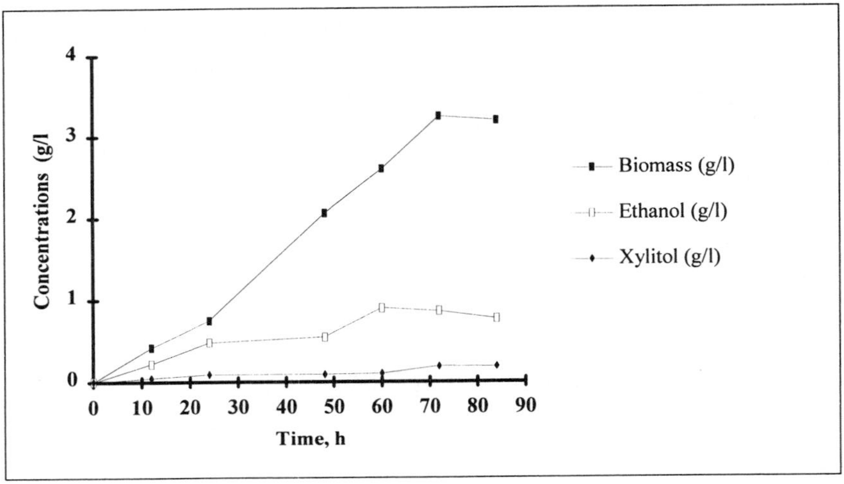

Fig. 3. Direct fermentation of oil palm waste xylose hydrolysate with *C. tropicalis* under optimum conditions (pH 5.5, temperature of 30°C and agitation at 200 rpm)

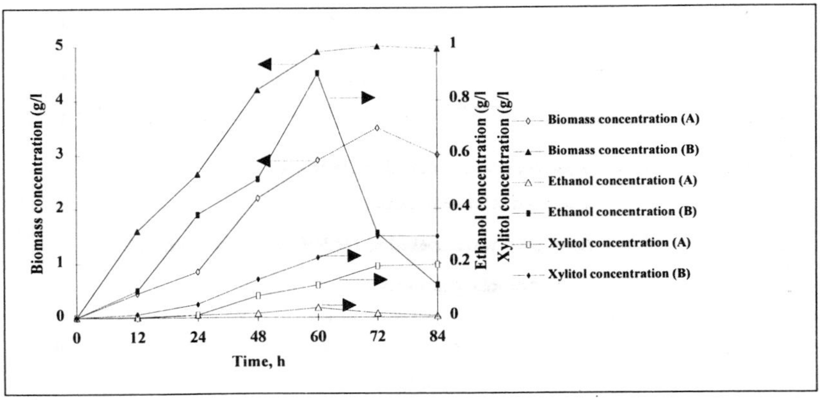

Fig. 4. Comparison of the growth of *C. tropicalis* and the production of ethanol and xylitol between the fermentation of the crude oil palm waste prehydrolysate (A) with and (B) without addition of growth factors (optimum conditions: pH 5.5, temperature of 30°C and agitation at 200 rpm)

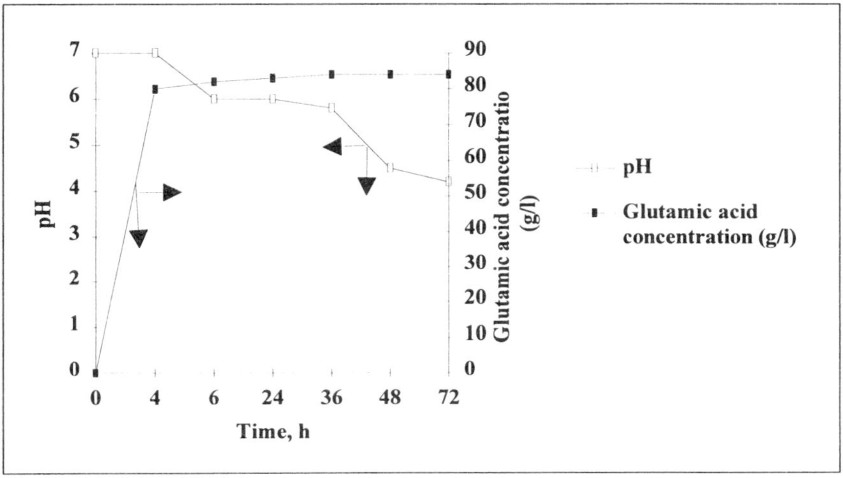

Fig. 5. Production of glutamic acid by fermentation of glucose hydrolysate by *B. lactofermentum* under optimum conditions (pH 7.5, temperature of 30°C and agitation at 200 rpm)

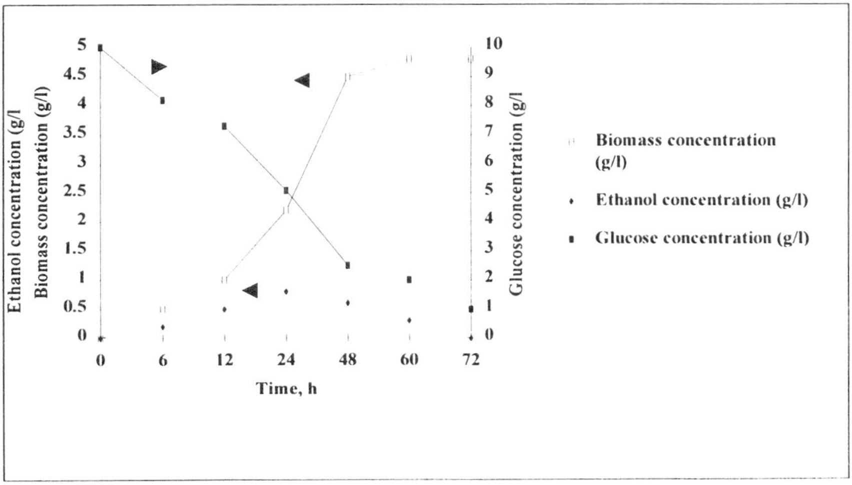

Fig. 6. Growth of *K. fragilis* and the production of ethanol during the fermentation of glucose hydrolysate under optimum conditions (pH 7.5, temperature of 30°C and agitation at 200 rpm)

FERMENTATIVE PRODUCTION OF NATURAL FOOD COLORANTS BY THE FUNGUS *MONASCUS*

Y.K. Lee*, D.C. Chen, B.L. Lim, H.S. Tay[+] and J. Chua[+]
Department of Microbiology, National University of Singapore, Kent Ridge, Singapore 0511
and [+]Food Biotechnology Centre, SISIR, 1 Science park Drive, Singapore 0511.

This study was an attempt to grow *Monascus* in submerged culture in stirred tank fermentors for the production of pigments. A high pigment producing *Monascus* mutant strain B683M was used. Highest pigment production was achieved under nitrogen limited growth condition. The type of N-source determined the ratio of yellow and red pigments produced, and it also determined whether the pigments were released into the culture medium or accumulated intracellularly. A nitrogen-limited fed batch culture system was the production system of choice. The pigments were thermostable at neutral and alkaline pH and degraded very slowly under fluorescent light.

INTRODUCTION

The red pigments of the fungus *Monascus* have been used as food colorants in the Orient for centuries. Lately, *Monascus* has been suggested as an avenue for natural food colorants to replace the conventionally use coal tordyes. The latter was implicated in carcinogenicity and may thus pose a potential threat to human health. Traditionally, the pigments are produced on steamed rice by solid state fermentation by *Monascus* (Steinkraus *et al*, 1978). In this paper, we report an attempt to produce *Monascus* pigments in submerged culture, and to incorporate modern concepts of microbial fermentation in the optimization of pigment production.

1. MATERIALS AND METHODS

Monascus sp B683M was a high pigment producing mutant of a locally isolated strain B683. For the physiological study, a chemically defined medium was used (amount in g/l) : glucose 50, MSG 6, $MgSO_4.7H_2O$ 8, KH_2PO_4 2.4, MOPS buffer 73.5. For C-limited culture, glucose was reduced to 16 g/l. In N-limited culture, the concentration of MSG was 2 g/l while the dissolved oxygen partial pressure of oxygen-limited cultures was maintained at 10%. In the mass cultivation of *Monascus*, glucose was replaced by tapioca starch and MSG by either protein A or protein B. In this case, the MOPS buffer was omitted. The *Monascus* was cultured in a 2.5 l stirred tank fermentor maintained at 28°C, 800 rpm, pH 5.7 and 20% pO_2 for the physiology studies. The fungus was cultured in 10 l and 600 l fermentors.

The dry weight (X) of the cultures was measured by oven-drying (80°C) the mycelia on filter paper for two days and then weighing. The specific growth rate of the cultures (μ) was calculated from the rate of increase in cell dry weight during the exponential growth phase. $\mu x = dx/dt$, where t = culture time.

The concentration of extracellular orange-red and yellow pigments was estimated from the relative absorbancy of filtrates at 500 nm and 400 nm respectively at 1 cm light path. To estimate the concentration of intracellular pigments, 10 ml of culture was spun down and washed with distilled water. The washed mycelia were then extracted by 50 ml pure methanol for 2 hours with continuous agitation. The extraction procedure was repeated when necessary. The concentration of red and yellow pigments was again estimated from the absorbancy of extracts at 400 nm and 500 nm respectively, taking the dilution factor into consideration. The two values (extracellular and intracellular) were added to give total pigment concentration. To calculate the specific rate of pigment production (Q_p), the rate of increase of total pigment concentration (dp/dt) during the exponential growth phase was divided by the dry weight in 1 l culture. $Q_p = (1/x).dp/dt$.

Residual glucose was determined using a YSI Biochemistry Analyser Model 2700 Select. The specific rate of glucose uptake (Q_g) by the cultures at exponential growth phase was calculated from the rate of depletion of glucose in culture medium (dg/dt). $Q_g = (1/x).dg/dt$.

Residual NH_4^+ was determined with an Ammonium Electrode 476 13 000J (Corning Ion Analyzer 150). The specific rate of NH_4^+ uptake (Q_n) in the cultures at exponential growth phase was calculated from the rate of depletion of NH_4^+ in the culture medium (dn/dt). $Q_n = (1/x).dn/dt$.

2. RESULTS

2.1 Kinetics of Growth and Pigment Production

As shown in Fig. 1, the production of both yellow and red pigments appeared to parallel the growth curve, and both pigment production and growth ended at around 140 h of culture time. The specific growth rate (μ), specific rate of pigment production (Qp), specific rate of glucose (Q_{glc}) and MSG (Q_{msg}) uptake during the exponential growth phase at between 20 and 120 h were estimated. The growth yield ($Y_{x/s}$) from the substrates and pigment yield ($Y_{P/X}$) were calculated accordingly.

2.2 Effect of Substrate Limitation

An attempt was made to compare pigment production at a particular growth rate (μ value chosen in this case was 0.004 - 0.005 h^{-1}) in different substrate limiting conditions, and the results are summarized in the Table 1.

TABLE 1. A comparison of the rate of pigment production, glucose uptake and MSG uptake during the exponential growth phase in oxygen; glucose- and MSG-limited cultures.

	O_2-Limited	C-Limited	N-Limited
Specific rate of pigment production, Q_p (OD units/mg biomass.h)	0.15	0.15	0.59
Specific glucose uptake rate,Q_{glc} (g/g biomass.h)	0.11	0.12	0.56
Specific MSG uptake rate, Q_{msg} (g/g biomass.h)	0.011	0.014	0.031

The Q_p of MSG-limited culture was about 4 times the production rate measured in oxygen- and glucose-limited *Monascus* cultures. Correspondingly, the Q_{glc} and Q_{msg} of MSG-limited culture were about 5 times and 2.5 times the values obtained from oxygen- and glucose-limited cultures. The Q_p, Q_{glc} and Q_{msg} of oxygen and glucose-limited cultures were comparable.

2.3 Intracellular vs Extracellular Pigments

The effect of N-source on the ratio of total intracellular and extracellular pigments is presented in Table 2.

TABLE 2 : The effect of N-source on the ratio of intracellular and extracellular pigments.

	MSG		Protein A		Protein B	
	Low	High	Low	High	Low	High
Intra/extra	18/1	1/1	3/2	1/6	20/1	1/10

It is clear that both the concentration of N-source in the culture medium and the type of N-source determined the release of pigments by *Monascus* cells. At low MSG or protein B concentrations, almost all the pigments produced by *Monascus* were retented in the mycelium. At high protein B concentration, about 90% of the pigments was found in the culture medium.

2.4 Red vs Yellow Pigments

The effect of dissolved oxygen partial pressure and N-source on the ratio of red (OD_{500}) and yellow (OD_{395}) pigments is presented in Table 3.

TABLE 3. The effect of dissolved oxygen partial pressure and N-source on the production of red and yellow pigments.

	Oxygen		MSG		Protein A		Protein B	
	Low	High	Low	High	Low	High	Low	High
Red/Yellow	1/2	1/1	1/3	1/1	1/9	1/4	1/1	1/1

More red pigment was produced in medium containing high concentration of MSG and protein B. High dissolved oxygen partial pressure also encouraged the production of red pigment. On the other hand, a low concentration of protein A resulted in production of mostly yellow pigment by the *Monascus*.

2.5 Stability of Pigments

Thermostability of the pigments was tested at boiling (100°C) and autoclave (121°C) temperatures (Table 4).

TABLE 4. Degradation of *Monascus* pigments at high temperature. The data represent % of residual pigments.

	pH5		pH7		pH9	
	Yellow	Red	Yellow	Red	Yellow	Red
100°C/30 min	48	57	100	100	99	100
121°C/20 min	26	23	66	84	66	90

Both red and yellow pigments were more thermostable at high pH value than at acidic pH value. Nonetheless, even at pH5, half of the pigments remained intact after boiling for 30 minutes.

2.6 Photostability of the Pigments

Direct exposure of *Monascus* pigments to sunlight at noon caused degradation of the pigments. However, fluorescent light had no detectable effect on the quality of the pigments for at least up to 2 hours of exposure.

TABLE 5. Photostability of *Monascus* pigments. The values represent % of residual pigments.

	Exposed to sunlight at noon (2000 $\mu E/m^2.s$) for 2 hours		Exposed to fluorescent light (30 $\mu E/m^2.s$) for 2 hours	
	Yellow	Red	Yellow	Red
5.0	46	35	100	100
7.0	51	37	100	100
9.0	48	39	100	100

3. DISCUSSION AND CONCLUSION

Monascus grew on simple sugar as well as on tapioca starch. Tapioca starch is a much cheaper carbon substract than rice which is traditionally used in *Monascus* red (Ang Kak) production (Steinkraus, 1978). Pigment production by *Monascus* cultures appeared to be growth-associated. While investigating the effects of substrate limitation, it was observed that the N-limited *Monascus* cultures took up MSG and glucose much faster than the O_2-limited and C-limited cultures, which may explain the higher rate of pigment production in the former. Through the manipulation of the type and concentration of the N-source, *Monascus* cultures could be induced to produce mostly yellow or red pigments, and to accumulate the pigments inside the mycelium or to release them into the culture medium.

The study suggests that *Monascus* pigments are suitable colorants for foods and beverages of neutral pH, or products which do not need to undergo long heat treatment and which are not exposed to direct sunlight during display or transportation, e.g. dairy products.

REFERENCE

Steinkraus KH, 1978. Handbook of Indigenous Fermented Foods. Marcel Dekker, New York. pp. 547-553.

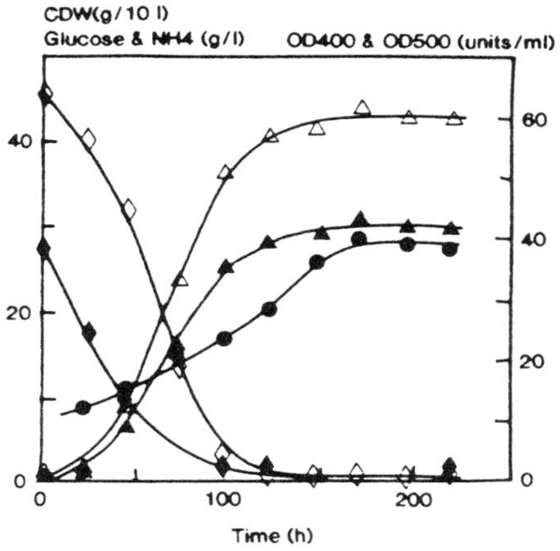

Figure 1. Fermentation kinetics of *Monascus* cultures in chemically defined medium. ●, cell dry weight; △, OD_{40}; ▲, OD_{50}; ◇, residual glucose; ◆, residual ammonium.

MICROBIAL CONVERSIONS OF AGRO-WASTE MATERIALS TO HIGH-VALUED OILS AND FATS.

Colin Ratledge
Department of Applied Biology, University of Hull, Hull, HU6 7RX, UK

The possibilities of producing microbial oils for edible purposes is reviewed. Although yeasts and moulds can produce a range of oils only a few have values above those of the major commodity oils and fats. Particular high-value added oils have been identified: a cocoa-butter like oil produced by *Cryptococcus curvatus* (=*Candida curvata*) and a number of fungal oils containing polyunsaturated fatty acids. Prospects of producing these from agro-wastes will be discussed as well as the prospects for their formation from existing plant-oils.

1. INTRODUCTION

Although all micro-organisms - bacteria, yeasts, moulds and algae - contain lipids, only a relatively small number accumulate lipid in such quantities to warrant consideration as potential sources of oils and fats. Such lipid-accumulating micro-organisms are termed *oleaginous*. Contents of lipids may be up to 80% of the cell dry weight and contents of about 50% lipids are not exceptional. The field of microbial lipids has been covered in a detailed two-volume, multi-authored treatise (1). Recent aspects of the commercial possibilities for using microbial lipids, termed *Single Cell Oils* (SCO), are covered in the volume edited by Kyle & Ratledge (2). Other recent reviews include those of Davies & Holdsworth (3), which looks at the prospects for producing oils using yeast technology, and Ratledge (4,5) which looks at the biotechnological potential for SCOs from all sources.

Physiologically, microbial lipids are produced as reserve, or storage, materials being synthesized when cells have exhausted a nutrient from their growth environment, but still are provided with a carbon source (see Fig. 1). The accumulated lipid often parallels the composition of oils and fats found in plants: i.e. is predominantly composed of triacylglycerols with the same range of fatty acyl substituents. The commercial drive is to identify those micro-organisms that produce the most valuable forms of lipid as it is obviously more profitable to produce a high priced oil than a lower priced one.

As it takes five or more tonnes of carbohydrate substrate to produce 1 tonne of oil in a micro-organism, the cost of the substrate is of critical importance if the overall costs of production are to be kept as low as possible in order to complete with existing sources of oil. Fortunately there are many agricultural wastes that may be suitable for upgrading in this way.

2. YEASTS

A number of yeast species are known that can accumulate more than 25% of their cell mass as extractable lipid (5). The lipid is predominanly composed of triacylglycerols and therefore represents a potential alternative to existing plant oils. The lipid contents and fatty acid profiles of the oleaginous yeasts are given in Table 1.

TABLE 1. Lipid contents and fatty acid profiles of oleaginous yeasts (from refs 1 + 5)

Yeast species	Maximum lipid Content (% w/w)	Relative % (w/w) of major fatty acyl residues 16:0	16:1	18.0	18.1	18.2	18.3†	Others
**Candida curvata* D	58	32	-	15	44	8	-	
**Candida curvata R*	51	31	-	12	15	6		
Candida didensii	37	19	3	5	45	17	5	18:4(1%)
Candida sp. 107	42	44	5	8	31	9	1	
Cryptococcus albidus var *albidus*	65	12	1	3	73	12	-	
Cryptococcus albidus var *albidus*	65	16	tr	3	56	-	3	21:0(7%) 22:0(12%)
Cryptococcus laurentii	32	25	1	8	49	17	1	
Endomycopsis magnusii	28	17	19	1	36	25	-	
Hansenula saturnus	28	16	16	-	45	16	5	
Lipomyces lipofer	64	37	4	7	48	3	-	
Lipomyces starkeyi	63	34	6	5	51	3		
Lipomyces tetrasporus	67	31	4	15	43	6	1	
Rhodosporidium toruloides	66	18	3	3	66	-	-	22:0(3%) 24:0(6%)
Rhodotorula glutinis	72	37	1	3	47	8	-	
Rhodotorula graminis	36	30	2	12	36	15	4	
Trichosporon cutaneum	45	30	-	13	46	11	-	
Trichosporon pullulans	65	15	-	2	57	24	1	
Yarrowia lipolytica	36	11	6	1	28	51	1	

* Also known as *Apiotrichum curvatum* but now called *Cryptococcus curvatus* (ref. 7).
† α-linolenic acid, 18:3(9,12,15).

Although there is some current interest in the feeding of yeasts to fish larvae mainly as a source of proteins and vitamins (6), these experiments have used almost exclusively *Saccharomyces cerevisiae* which not only has a low lipid content but also does not contain any di- or poly-unsaturated fatty acids considered to be nutritionally important for the fish larvae. As the energy-content of high-lipid containing yeasts is considerably greater than a protein-rich yeast, and as many such yeasts contain substantial proportion of 18:2 plus 18:3 fatty acids (see Table 1), it is clear that these yeasts would represent better sources of energy and essential fatty acids than the other non-oleaginous yeasts. However, the cost of producing such yeasts is likely to be far above that of producing *S. cerevisiae* which can be easily obtained from industrial processes producing baker's yeast, that it seems unlikely that the oleaginous yeasts, although nutritionally superior could be produced cost effectively.

Prospects of using yeast oils commercially therefore probably reside in identifying a higher valued lipid product than just one needed for fish feed. Such a product has been considered to be a substitute for cocoa butter or 'cocoa-butter equivalent' (CBE).

Of particular interest in this area is the yeast *Cryptococcus curvatus* (formerly known as *Candida curvata* and *Apiotrichum curvatum* and now re-classified once more - see ref. 7). This yeast grows exceptionally well on lactose which is the principal carbohydrate of whey, the waste material arising from cheese manufacture. Indeed, the yeast was originally isolated from a cheese dairy processing plant (8). Deproteinized whey is therefore the substrate of choice for this process (8,9,10) though the yeast will perform equally well on molasses (11) or a number of other substrates containing glucose, galactose, fructose, xylose etc either singly or as mixtures (12,13). It has also been successfully grown on the juice from waste bananas (14).

The major results (see Table 2) have been achieved by Julian Davies in New Zealand (3,10,15) and a group, led by Henk Smit, working in the Netherlands (16,17).

TABLE 2. Fatty acyl composition of triacylglycerols from *Cryptococcus curvatus* wild type (WT) strain, an unsaturated fatty acid auxotrophic mutant (Ufa 33), a revertant mutant (R22. 72) and a hybrid derived from Ufa 33 (F33. 10) compared to cocoa butter fatty acids (from ref. 16,17)

	Relative % (w/w) of major fatty acyl groups					
Yeast strain	16:0	18:0	18:1	18:2	18:3	24:0
WT	17	12	55	8	2	1
Ufa 33[a]	20	50	6	11	4	4
R22. 72	16	43	27	7	1	2
F33. 10	24	31	30	6	-	4
Cocoa butter	23-30	32-37	30-37	2-4	-	trace

[a] Grown with 0.2g oleic acid/l.

The key feature of the yeast fat is its very high content of stearic acid (18:0) thus making it similar in physical characteristics to cocoa butter used in chocolate manufacture. Trials of the yeast CBE admixed with cocoa butter have produced very acceptable grades of chocolate and the yeast CBE could certainly be used in confectionery-grade chocolate, if not in chocolate itself, if local regulations would permit this. However the current world price of cocoa butter itself is now only $US

3000/tonne having been over \$8000/tonne in the mid-1980s. This low price has forced the price of a substitute to less than \$US 2000/tonne. As the New Zealand yeast CBE process was calculated on the basis of having to build a new fermentation plant, and without financial credit for removing a waste material (the whey), the economics of the process are now not considered favourable (15).

Clearly such a process could only be economical if a large, continuous supply of carbohydrate-rich feedstock could be identified and, moreover, would probably have to be used in existing fermentation plant (~200 to 1000 m^3 capacity). The tonnage of feedstock required would have to be of the order of 10000 tonnes/year to make the process cost-effective. This then places considerable onus on ensuring that an adequate market for the product exists.

Without doubt, the technology exists for producing a CBE from *Crypt. curvatus*: what is needed is a market for the product at the price that would be attractive for the producers. For countries like Malaysia and its neighbours, the prospects for this process must be regarded as minimal as it is clearly a product that would complete against either indigenous cocoa (cacao) production or palm oil fractions that are currently the main source of CBEs. Thus, even if there were major sources of fermentable carbohydrate available in Malaysia and other tropical countries, it seems unlikely that a yeast-CBE would be a target product. Consideration could therefore be given to using any such feedstocks for the production of polyunsaturated fatty acids that are now in increasing demand but would, though, have to be produced using fungal technology.

3. FUNGI

Fungi produce a much wider range of lipid products than yeasts. A larger number of oleaginous species are known (5) some of which are shown in Table 3 together with the fatty acid profiles of the lipids.

Several fungi are now under active investigation for the production of specific polyunsaturated fatty acids (PUFAs) that have nutritional and dietetic interest. The principal PUFAs are γ-linolenic acid [18:3(6,9,12)], arachidonic acid [20:4(5,8,11,14)], eicosapentaenoic acid [20:5(5,8,11,14,17)] and docosahexaenoic acid [22:6(4,7,10,13,16,19)]. These are referred to, respectively, as GLA, ARA, EPA and DHA. The demand for oils containing these fatty acids is mainly in the field of dietary supplements and the total world volumes required are probably all less than 1000 tonnes/year. EPA and DHA are usually obtained together from fish oil and as such are not particularly expensive though there is difficulty in obtaining each acid in a pure form. However, there is as yet no particularly large demand for each acid to be produced relatively free from the other.

Until recently the main drive has been to produce GLA-rich oils using fungal technology as a rival to the plant sources of evening primrose seed oil and borage seed oil (see Table 4). Both these latter oils enjoy considerable sales as over-the-counter dietary supplements in the UK and Europe. The market potential for ARA and DHA probably are as essential fatty acids being added to infant milk formulations as cow's milk used as a mother's milk substitute does not contain these fatty acids (nor does it contain GLA). Mother's milk, on the other hand, contains a range of PUFAs including GLA, ARA, EPA and DHA (18) which are considered nutritionally important for development of newly-born babies. Whether fungal oils containing these PUFAs would receive regulatory approval is, as yet, an open question but there is no *a priori* reason why they should not as the experienced gained in producing GLA-rich oils by the fungal route indicates that these oils are extremely safe and have no deleterious residues that would be harmful to any consumer - young or old.

Commercial processes for the fungal production of GLA have been run in the UK and Japan (4). The cost of the feedstock is not considered as critical as that for yeast-CBE though, because the annual production from a given site would probably not exceed 100 tonnes/year, the amount of feedstock required would not be excessive. These specialized fungal SCO processes would therefore be limited to the upgrading of small amounts of agro-waste material such as starch-wastes from rice, sago or even cassava processing.

TABLE 3. Lipid contents and fatty acids of some selected moulds (from refs 1 & 5)

Organism	Lipid fraction (% cell dry wt)	Relative major fatty acyl groups						
		14:0	16:0	18:0	18:1	18:2	18:3	Others
Phycomycetes								
Entomophthorales								
Conidiobolus nanodes	26	1	23	15	25	1	4*	20:1(13%) 22:1(8%) 20:4(4%)
Entomophthora coronata	43	31	9	2	14	2	1*	12:0(40%)
Entomophthora obscura	34	8	37	7	4	tr	tr	12:0(41%)
Mucorales								
Absidia corymbifera	27	1	24	7	46	8	10*	-
Cunninghamella japonica	60	tr	16	14	48	14	8*	-
Mortierella isabellina	86	1	29	3	55	3	3*	-
Rhizopus arrhizus	57	19	18	6	22	10	12*	-
Mucor alpina-peyron	38	10	15	7	30	9	1*	20:0(8%) 20:3(6%) 20:4(5%)
Ascomycetes								
Aspergillus terreus	57	2	23	tr	14	40	21	-
Fusarium oxysporum	34	tr	17	8	2	46	5	-
Pellicularia practicola	39	tr	8	2	11	72	2	-
Hyphomycetes								
Cladosporium herbarum	49	tr	31	12	35	18	1	-
Ustilaginales								
Tolyposporium ehrenbergii	41	1	7	5	81	2	-	-
Clavicipitacae								
Claviceps purpurea	60	tr	23	2	19	8	-	12-HO-18:1(42%)

* = γ - linolenic acid (18:3ω6)

Fungal species are known that will produce all the above mentioned polyunsaturated fatty acids as individual entities (see Table 5 and refs. 4 and 5). It therefore remains for interested parties to identify the market niche they wish to attack and then devise the best biotechnological strategy to produce the requisite oil. Cost-effective processes will depend more on marketing skills than on development of the process which, scientifically, is relatively easy.

TABLE 4. Fatty acid profiles of two commercial fungal oil products grown on glucose compared with evening primrose oil, borage oil and blackcurrant oil containing γ- linolenic acid (from refs 1, 4 and 5)

	*Mucor circinelloides**	*Mortierella isabellina†*	Evening Primrose	Borage	Blackcurrant
Oil Content (% w/w)	20	ND	16	30	30
Fatty acid	Relative percentage of fatty acids in oil (%w/w)				
16:0	22-25	27	6-10	9-13	6
16:1	0.5-1.5	1	-	-	-
18:0	5-8	6	1.5-3.5	3-5	1
18:1	38-41	44	6-12	15-17	10
18:2	10-12	12	65-75	37-41	48
γ-18:3	15-18	8	8-12	19-25	17
∝-18:3	0.2	-	0.2	0.5	13
20:1	-	0.4	0	4.5	-
22:1	-	0.2	-	2.5	-
24.0	-	-	-	1.5	-

ND = not disclosed.

* Production organism used by J & E Sturge, Selby, UK; 1985-1990.

† Production organism used by Idemitzu Ltd, Japan; 1988 - 1992.

TABLE 5. Moulds as sources of polyunsaturated fatty acids (from ref. 5)

Fatty acid	Producing organism	% fatty acid in oil
γ-Linolenic acid [18:3(6,9,12)]	*Mucor circinelloides*	18
	Mortierella isabellina	8
Dihomo-γ-linolenic acid [20:3(8,11,14)]	*Mortierella alpina* (+ sesame oil)	29
Eicosatrienoic acid 'Mead Acid' [20:3(5,8,11)]	*Mortierella alpina* (mutant strain)	15
Arachidonic acid [20:4(5,8,11,14)]	*Mortierella alpina* (ATCC 32221 and 25-4)	65 to 70
Eicosapentaenoic acid [20:5(5,8,11,14,17)]	*Mortierella alpina* (20-17)	8
	Mortierella elongata (NRRL 5513)	20
Docosahexaenoic acid [22:6(4,7,10,13,16,19)]	*Thraustochytrium aureum*	50

4. BIOTRANSFORMATION OF PLANT OILS TO HIGHER VALUED OILS.

Plant oils being relatively cheap commodities would appear, at least on paper, to be the ideal substrate to present to a micro-organism for its conversion to an improved or higher-valued product. Unfortunately most micro-organisms when presented with an oil or fatty acid on which to grow do not carry out either elongation or desaturation of it. The presented oil usually satisfies the requirement of that cell for fatty acyl groups and elongation enzymes and desaturases are consequently repressed when yeasts and moulds are grown on such materials. Table 6 shows some recent results using *Mucor circinelloides* grown on triolein and on coconut oil. It is clear that with both substrates there is very little modification of their fatty acids, even with coconut oil there is very little evidence of elongation of the short-chain lauric acid (12:0).

TABLE 6. Fatty acids produced by *Mucor circinelloides* after growth on triolein and coconut oil compared to the original fatty acid composition

	Rel. & (w/w) fatty acids						
	12:0	14:0	16:0	18:0	18:1	18:2	18:3*
Triolein substrate	-	-	10	5	66	17	2
fungal oil	-	-	17	1	68	10	1
Coconut oil substrate	54	20	12	3	9	2	-
fungal oil	44	21	13	4	13	2	-

* = α-linolenic acid - 18:3(9,12,15).

However Shimizu *et al.* (19) have shown that there are some exceptions to this. *Mortierella alpina* was able to convert and utilize various oils and could apparently elongate and desaturate linseed oil that contained α-linolenic acid, 18:3(9,12,15) into EPA-20:5(5,8,11,14,17). Not all species of *Mortierella* showed this ability for transformation and therefore as conversions with the above mentioned strain of *M. alpina* were not excessive, it is clear that progress in this direction must lie in being able to prevent the repression of the elongases and desaturases by genetic modification.

Fatty acids may though undergo other biotransformation reactions. Conversions of them to long chain dicarboxylic acids and to ω-hydroxyfatty acids are well-known (5,20) as is the formation of various glycolipids that contain substituted fatty acids. These conversions all require, however, very cheap sources of fatty acids but in Malaysia these materials are probably available at the lowest price anywhere in the world. It would therefore be prudent for Malayan biotechnologists to examine the prospects of producing some of the unusual fatty acids or glycolipids.

Like all products, however, a likely market for them has to be identified before any research programme is undertaken.

REFERENCES

1. Ratledge, C. and Wilkinson, S.G., editors (1988 and 1989). *Microbial Lipids*, vols. 1 & 2. Academic Press, London.
2. Kyle, D.J. and Ratledge, C., editors (1992). *Industrial Applications of Single Cell Oils*. American Oil Chemists' Society.
3. Davies, R.J. and Holdsworth, J.E. (1992). *Adv. Appl. Lipid Res.* **1**, 119-159.
4. Ratledge, C. (1993) *Trends in Biotechnol.* **11**, 278-284.
5. Ratledge, C. (1993) In *Improved and Alternative Sources of Lipids*. Ed. B.S. Kamel, Blackie Academic Publishers, Glasgow, UK, pp. 235-291.
6. Watanabe, T., Ohita, M., Kitajima, C. and Fujita, S. (1982). *Bull. Jap. Soc. Sci. Fish.* **48**, 1775-1782.
7. Barnett, J.A., Payne, R.W. and Yarrow, D. (1990). *Yeasts*: characteristics and identification. 2nd edn. Cambridge University Press, Cambridge, UK.
8. Moon, N.J., Hammond, E.G. and Glatz, B.A. (1978). *J. Dairy Sci.* **61**, 1537-1547.
9. Ykema, A., Verbree, E.C., Kater, M.M. and Smit, H. (1988). *Appl. Microbiol. Biotechnol.* **29**, 211-218.
10. Davies, R.J. (1988). In *Single Cell Oil*. Ed. R.S. Moreton, Longman, Harlow, UK, pp. 99-145.
11. Bednarski, W., Leman, J. and Tomasik, J. (1986). *Agric. Wastes* **18**, 19-26.
12. Evans, C.T. and Ratledge, C. (1983) *Lipids* **18**, 623-629.
13. Heredia, L. and Ratledge, C. (1988) *Biotech. Lett.* **10**, 25-30.
14. Vega, E.Z., Glatz, B.A. and Hammond, E.G. (1988). *Appl. Environ. Microbiol.* **54**, 748-752.
15. Davies, R.J. (1992) *Lipid Technol.* **4**, 6-13.
16. Ykema, A., Verbree, E.C., Kater, M.M. and Smit, H. (1988). *Appl. Microbiol. Biotechnol.* **29**, 211-218.
17. Ykema, A. *et al.* (1990) *Appl. Microbiol. Biotechnol.* **33**, 176-182.
18. Gibson, R.A. and Kneebone, G.M. (1981) *Amer. J. Clin. Nutr.* **34**, 252-257.
19. Shimizu, S. *et al.* (1989) *J. Amer. Oil Chem. Soc.* **66**, 342-347.
20. Ratledge, C. (1993) In *Biochemistry of Microbial Degradation* (ed. Ratledge, C.) Kluwer Academic Publ., Dordrecht, pp. 89-141.

Fig 1. Idealized lipid accumulation pattern in an oleaginous yeast or mould growing in a medium with a high C:N ratio.

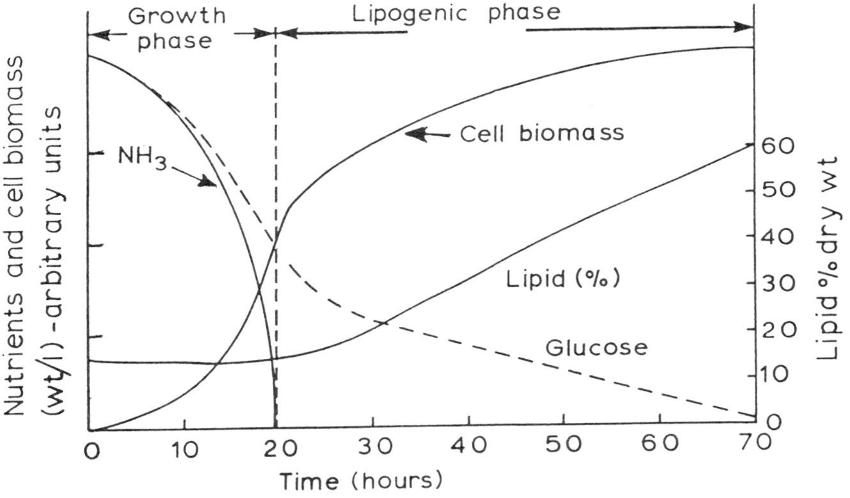

PERFORMANCE EVALUATION AND MODIFICATIONS OF A UASB REACTOR TREATING SUGAR INDUSTRY EFFLUENT: A CASE STUDY

Grasius M.G., Iyengar L[1], Singh H.B.[2], Venkobachar C.
Environmental Engineering Laboratory
Indian Institute of Technology, Kanpur - 208 016, India

1. Department of Chemistry, IIT, Kanpur, India
2. Director, Hydro Air Tectonics, Bombay, India

The efficiency of the Effluent Treatment Plant (ETP) of a cane sugar mill in North India consisting of Upflow Anaerobic Sludge Blanket (UASB) reactor along with primary treatment facilities was evaluated. A low COD removal efficiency of 37.5% by the UASB warranted further detailed study of this unit. Low concentration of Volatile Suspended Solids (VSS) and short circuiting of the flow were major shortcomings for which remedial measures were suggested. Reseeding was done to increase the VSS content of the UASB before the next crushing season. During the first 30 days of operation, the soluble COD removal efficiency of UASB has reached 66%.

1. INTRODUCTION

Anaerobic biomethanation is an emerging field of biotechnology, which is now increasingly being used for environmental protection and recovery of renewable resources. This renewed popularity for the process is due to the development of several high rate reactor configurations based on a better comprehension of the process intricacies. It is a multistaged process by which the organic compounds in the waste are biologically converted to CH_4 and CO_2 in the absence of molecular oxygen. For the successful completion of the process, coordinate participation of at least three groups of bacteria are required (1). The major biochemical transformations brought about by these microbial consortium are hydrolysis and fermentation of polymers to volatile fatty acids and alcohols, conversion of these intermediates to acetate and H_2, and finally formation of methane from acetate as well as from CO_2 and H_2. A simultaneous transfer of electrons from Obligate H_2 Producing Acetogens (OHPA) to H_2 utilizing methanogens during the degradation of intermediates, termed as interspecies electron or H_2 transfer, is considered as a crucial step in the process. Also, the conversion of acetate to CH_4 is identified as the rate limiting step in stabilization of soluble substrates.

The high rate reactors are essentially designed to retain high concentrations of microorganisms in the reactor, thus achieving a high organic loading rate as well as treatment efficiency (2). Secondly, the 'juxtapositioning' of OHPA and H_2 utilizing methanogens in these 'Retained Biomass System' helps in achieving an enhanced electron or H_2 transfer resulting in an efficient conversion of organic matter (3). Among the various novel reactor configurations, the Upflow Anaerobic Sludge Blanket (UASB) is the most popular and is being used extensively for treatment of domestic and industrial wastewaters (4).

This paper presents a case study of an Effluent Treatment Plant (ETP) of a cane sugar mill. This industry located in North India is more than 60 years old. It has a crushing capacity of 1000 tonnes of sugar cane/day and generates a waste discharge of 350 m^3 /day. The factory was having treatment facilities, essentially consisting of Primary Sedimentation Tank (PST) and Trickling Filter (TF) to which a UASB reactor and a polishing unit (Extended Aeration) were recently added (Fig.1). These units were designed, constructed and commisioned by M/s Hydro Air Tectonics, New Bombay, India.

The objectives of this study were to evaluate the performance of ETP as a whole and the UASB in particular and to suggest possible modifications with subsequent evaluation.

2. METHODOLOGY

Samples were collected at two times from the ETP, one at the middle and another at the end of the just previous crushing season (November 1992 to June 1993). Reseeding was done (details given later) to increase the VSS content of the reactor just before the next crushing season. After a start-up period of a 30 days, samples were collected to study the effect of this. The sampling points are indicated in Fig.1.

2.1 Study on Flow Fluctuations and Sample Compositing

Flow, temperature and pH were measured at 2 h interval at the influent and effluent V-notches located in the plant for a period of 24 h. The samples were then composited and analyzed for total/soluble solids.

2.2 Study on Performance of Individual Units

Two samples each were collected at an interval of 12 h from the sump well (inlet to UASB) and outlet of UASB, and were analyzed for various parameters.

2.3 Study on UASB Reactor

This unit made of mild steel is of 6 m diameter and 4.5 m height. It has a reaction liquid volume of 113 m^3 excluding settler volume. The waste is distributed from bottom through nozzles fixed on distribution pipes.

To study the reactor performance, the mixed liquor was drawn from four different heights of the reactor. The samples were analyzed for soluble COD, TSS, VSS, and Volatile Fatty Acids (VFA). The settled sludge from these samples was mixed together and subjected to specific methanogenic activity test as per the method given by Valcke and Verstraete (5). Direct titration method given by DeLallo and Albertson (6) was used for VFA analysis. All other analyses were done as per Standard Methods (7).

To study the hydraulic efficiency of the unit, a slug dose of tracer (NaCl) was injected to the suction end of the intake pump. Tracer concentrations in the effluent of UASB were measured at 30 minutes intervals, for 12 h.

3. RESULTS AND DISCUSSIONS

The results of the analysis of samples from ETP taken at two times during November 1992 to June 1993 are given in Table 1 and Table 2. The average inflow to the ETP was 350 m^3/day during the middle of the crushing season.

TABLE 1 - Analysis of Samples Collected at Middle of Crushing Season from ETP

Sample Desig-nation	Details	COD, mg/L Total	Soluble	Total Solids mg/L	Removal eff. (Soluble COD) Unit	Percent	pH
I	Influent to ETP	2760	2360	376	-	-	7.3
S	Sumpwell influent to UASB	1000	800	339	PST	66	6.5
B	Effluent from UASB	860	560	41	UASB	30	6.5
E	Effluent from ETP	680	480	207	TF & EA	14	7.4
					Over-all	80	-

TABLE 2 - Analysis of Samples Collected at End of Crushing Season from ETP

Sample Desig-nation	Details	COD, mg/L Total	Soluble	Solids Total mg/L	Soluble	Removal (Soluble COD) Unit	Percent	pH
I	Influent to ETP	640	520	460	300	-	-	6.6
S	Sumpwell influent to UASB	224	160	160	80	PST	69	6.7
B	Effluent from UASB	160	100	220	120	UASB	37.5	6.9
E	Effluent from ETP	96	80	170	130	TF & EA	20	7.8
						Over-all	85	-

TF - Trickling Filter : EA - Extended Aeration

The influent to ETP had a total COD of 2760 mg/L and a soluble COD of 2360 mg/L. Total COD of the final effluent was 680 mg/L while the soluble COD was 480 mg/L. The efficiency in soluble COD reduction was nearly 80%. However the UASB reactor showed only a marginal COD removal of 30%. A significant reduction in soluble COD was observed in the PST. This is attributable to the very active anaerobic decomposition taking place in the PST which contains considerable quantity of settled sludge. The combined efficiency of Trickling Filter and Extended Aeration was only 14%.

When similar studies were repeated at the end of the crushing season (May, 1993), slight improvement in efficiency was observed in these units. The overall efficiency increased to 85% while UASB efficiency increased to 37.5% (Table 2).

3.1 Performance of UASB Reactor

Samples from different depths of UASB were drawn and subjected to detailed diagnostic tests and the results are reported in Table 3 and Table 4. The UASB unit was operating at 7.75 h Hydraulic Retention Time (HRT). Though the specific methanogenic activity of the digester sludge was reasonable (60 and 90 ml CH_4/g.VSS.day), average VSS concentrations were only 0.7 and 0.5 g/L respectively for the two sampling periods. The VSS concentration in an efficiently working UASB is reported to be around 20-40 g/L (8). So the very low VSS concentration in the reactor is one of the major reasons for poor performance of the UASB. Further, the VSS profile along the height of the reactor showed a higher biomass content at the top. This indicates that the gas/solid/liquid separation system is not permitting the washout of fluffy and light biomass which is essential for the development of active granular sludge with good settleability.

TABLE 3 - Analysis of Samples Collected at Middle of Crushing Season from UASB

S. No	Height of sampling from Reactor Bottom (m)	Soluble COD (mg/L)	TSS mg/L	VSS mg/L	Methanogenesis activity (mL CH_4 /g VSS day)
1	3.5	440	3065	1985	
2	2.5	480	1023	645	60
3	1.5	440	65	37	
4	0.5	600	335	190	

TABLE 4 - Analysis of Sample Collected at End of Crushing Season from UASB Reactor

S. No.	Height of sampling from Reactor Bottom (m)	Soluble COD (mg/L)	TSS mg/L	VSS mg/L	Methanogenesis activity (mL CH_4/g VSS day)
1	3.5	120	1240	610	
2	2.5	133	830	410	90
3	1.5	107	810	390	
4	0.5	147	1150	610	

The data obtain from tracer studies was analyzed to evaluate the hydraulic efficiency of the unit (9). A central tendency value namely, ta/T in which, ta is the time corresponding to the centroid of the area confined by flow curve and T the theoretical detention time, was determined. This ratio was obtained as 0.52 which indicates the presence of considerable dead spaces and thereby short circuiting in the reactor. The lack of uniformity in wastewater distribution at the bottom of the reactors appears to be the major contributing factor for this short circuiting.

Based on the above observations the following remedial measures were suggested:

1. Reseeding of the reactor to achieve a VSS concentration of 10-15 g/L. The actively digesting sludge from the PST could be used for this purpose. The hydraulic loading to the UASB should only be increased gradually, reaching the full space loading in about 30 to 40 days.

2. Modification of the gas/solid/liquid separation system by providing sufficient aperture between the deflector beam and gas collection cone. This is expected to achieve continuous removal of lighter sludge fractions from the reactor permitting the retention of heavier sludge ingredients.

3. Modification of the waste flow distribution system to ensure uniform distribution of incoming flow at the bottom of the reactor. This is expected to reduce the percentage of dead spaces and thus short circuiting.

3.2 Reseeding, Start-Up and Performance Evaluation of the UASB Reactor

The two latter remedial measures requiring structural modifications, are presently not undertaken. However, the reseeding of the unit was done just before the next crushing season. The sludge from the PST which has a specific methanogenic activity of 25 ml CH_4/g.VSS.day was used for this. The reactor was charged with sufficient quantity of sludge to achieve VSS concentration of 12 g/L. By middle of November 1993 the reactor was loaded with wastewater at an HRT of 12 h. After 7 days of operation, the flow rate was gradually increased to achieve an HRT of 8 h by 21st day. The influent and effluent samples of UASB were collected after 30 days of operation. These samples were analysed for total and soluble COD. Samples from four heights of the reactor were collected, composited and subjected to TSS and VSS determination.

The total COD of influent and effluents from the UASB were 1200 mg/L and 420 mg/L respectively while the soluble COD were 960 mg/L and 320 mg/L. The efficiency in soluble COD removal was 66.67%. The average TSS and VSS contents in the reactor were 27 g/L and 13 g/L respectively. The space loading to the reactor to this stage was 3.6 g COD/L. day which corresponds to a sludge loading of 0.28 g COD/g VSS. day. Even though granulation was not observed in the reactor, the sludge was found to have good settling characteristics. Continued operation of the reactor at the HRT of 8 h is expected to increase the efficiency further.

4. CONCLUSIONS

1. The overall efficiency of the ETP was 85%.

2. The contribution of recently incorporated UASB treatment unit to the overall COD reduction was only marginal. The major reason for this is the low VSS content in the reactor. Presence of dead spaces and short circuiting of flow in the reactor also may be contributing to this.

3. After reseeding and initial start-up operation of 30 days the soluble COD removal efficiency of UASB was 66.67%.

4. The gas/solid/liquid separation system is to be modified to permit washout of light fractions of sludge.

5. The wastewater distribution system of the UASB reactor is to be modified.

5. REFERENCES

1. Gujer, W. and Zehnder, A.J.B., 1983, Water Sci. Technol., 15, 127.

2. Iza, J., Colleran, E., Paris, J.M. and Wu. W.M., 1991, Water Sci. Technol., 24,1.

3. Thile, J.H., Chartrain, M. and Zeikus, J.G., 1988, Appl. Env. Microbiol., 54, 10.

4. Pol, L.H. and Lettinga, G., 1986, Water Sci. Technol., 12, 41.

5. Valcke, D. and Verstraete, W., 1983, J. Water Poll. Cont. Fed., 55, 1191.

6. DeLallo, R. and Albertson, O.F., 1961, J. Water Poll. Cont. Fed., 33, 356.

7. Standard Methods for the Examination of Water and Wastewater, 17th Ed., 1989, APHA, AWWA, WPCF.

8. Stronach, S.M., Rudd, T. and Lester, J.N., 1986, Anaerobic Digestion Process in Industrial Wastewater Treatment, Springer - Verlag, Berlin.

9. Rebhun, M. and Argaman, Y., 1965, Proc. ASCE, 91, 37.

6. ACKNOWLEDGEMENTS

M/s Hydro Air Tectonics, New Bombay, India are acknowledged for sponsoring one of the authors (Grasius M.G.) to the symposium.

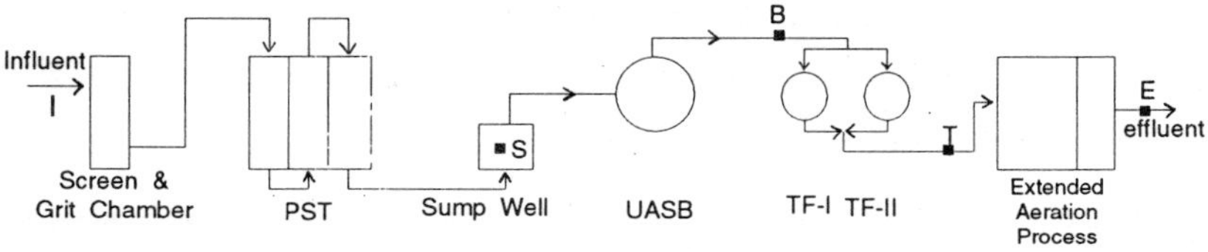

I,S,B,T and E: Sampling Points

Layout of the Effluent Treatment Plant of the Sugar Industry

GROWTH AND PRODUCT FORMATION OF *ANKISTRODESMUS CONVOLUTUS* IN AN AIR-LIFT FERMENTER

W.L. Chu, S.M. Phang and S.H. Goh

Institute of Advanced Studies, University of Malaya, 59100 Kuala Lumpur, Malaysia

Abstract -Four batches of *Ankistrodesmus convolutus* cultures in an airlift fermenter were investigated in terms of growth and biochemical composition. The following culture conditions were employed: (1) unbuffered at low light intensity, (2) buffered at low light intensity, (3) buffered at high light intensity and (4) buffered at low light intensity and aerated with air containing 5% CO_2. The highest biomass (308 mg dry weight/L) was attained by the culture aerated with air containing 5% CO_2. Cells grown at high light intensity contained more lipids (26.0% dry weight) than at other culture conditions studied. No significant variation in fatty acid composition (predominantly 18:3n-3) was observed. The algal cells grown at high light intensity had less chlorophyll-a content than those grown at low light intensity. Contents of other pigments did not differ significantly.

1. INTRODUCTION

Microalgae are potential sources of a wide range of useful natural products. The products which are commercially important include specialty lipids (eg. arachidonic acid, eicosapentaenoic acid and docosahexaenoic acid) and pigments such as carotenoids and phycobiliproteins (1).

Mass culturing of microalgae is vital in generating sufficient biomass for the extraction of various products. In parallel with the advancement of microalgal biotechnology, various mass-culture systems have been developed. Mass culture systems of microalgae include both open (raceway pond) and closed systems (phototubular bioreactor and air-lift fermenter). Closed systems are more suitable than open systems for physiological and productivity studies which require strictly sterile conditions.

Air-lift bioreactors are advantageous for the culturing of microalgae because they offer simple and effective mixing with no moving parts, high gas absorption efficiency and good heat transfer characteristics (2).

A. convolutus is a fast growing-alga producing appreciable amounts of carotenoids and polyunsaturated fatty acids (3). This paper examines the growth trend and biochemical composition of *A. convolutus* cultured in an air-lift fermenter.

2. MATERIALS AND METHODS

2.1 Culture Conditions

Ankistrodesmus convolutus Corda (isolate No. 101) used in the present study was isolated from a freshwater pond and deposited at the Microalgal Culture Collection Centre at the Institute of Advanced Studies, University of Malaya. The alga was grown in an airlift fermenter (cap. 6.5 L) using Bold's Basal Medium, BBM (4). Inocula (10%) were obtained from exponential phase cultures. The optical densities at 620 nm of the inocula were standardised at 0.2.

Four batches of cultures were grown in the air-lift fermenter under the following conditions: (1) unbuffered at low light intensity (31.9 $\mu Em^{-2}s^{-1}$), (2) buffered with 10 mM N-2-hydroxyethyl piperazine-1-ethanesulphonic acid (HEPES) at low light intensity (3) buffered at high light intensity (63.8 $\mu Em^{-2}s^{-1}$) and (4) buffered at low light intensity and aerated with air containing 5% CO_2. Batches 1, 2 and 3 were aerated with filter-sterilised air whilst batch 4 with air containing 5% CO_2 at 1,500 mL/min. Illumination was set on a 12 h light and 12 h dark cycle. Temperature was maintained at 28° C throughout the study.

2.2 Growth Monitoring

Growth was monitored daily by cell counting and OD measurements at 620 nm (OD_{620}).

2.3 Chemical Analyses

On reaching stationary phase, the cells were filtered (Whatman GF/C , 0.45 μm) for chemical analyses and dry weight determinations (100°C, 24 h).

Lipids were extracted in methanol-chloroform-water (2:1:0.8 v/v/v) and determined gravimetrically (5). Fatty acid methyl esters were analysed by gas chromatography after transesterification of the lipids in 1 N sodium methoxide at 60° C for 20 min (6). The gas chromatograph was equipped with a polar capillary column (DB 23). The following temperature program was used : 130°C (2 min) increased to 200°C (1 min) at a rate of 3° C/min then further increased to 230°C at 2°C/min and held for 5 min.

Proteins were extracted in 0.5 N NaOH (80°C, 20 min) and assayed according to the dye-binding method (7). Carbohydrates were determined using the phenol-sulphuric acid method (8) after extracting the cells with 2 N HCl (80°C, 20 min).

Pigments were extracted by sonication (2 min) in 100% acetone (HPLC grade). After centrifugation, the supernatants were removed, evaporated and redissolved in 1 mL of acetonitrile-methanol-acetone (56:40:4 v/v/v) before analysis by HPLC. The HPLC system consisted of a Rheodyne valve injection port (200 μL loop), a Shimadzu LC 7A pump and a Shimadzu M6A photodiode array detector. The pigments were resolved using a reversed phase column (C18) filled with 5 μ materials and a dimension of 300 X 39 mm.

The pigments were separated isocratically using mobile phase consisting of acetonitrile:methanol:acetone (56:40:4 v/v/v) at a flow rate of 1 mL/min. Identification and quantification of the pigments were based on comparison and calibration with authentic standards supplied by Sigma and Fluka Chemicals.

3. RESULTS AND DISCUSSION

3.1 Growth Trend

The culture aerated with air containing 5% CO_2 attained higher cell density and OD_{620} than other batches (Figures 1a and 1b). No marked difference was observed in growth trend among the other batches of cultures studied.

The highest biomass based on dry weight was attained by the culture aerated with air containing 5% CO_2 (Table 1). For comparison, the addition of CO_2 was reported to improve biomass production in *Tetraselmis suecica*, another green alga (9). The culture grown at high light intensity (63.8 $\mu Em^{-2}s^{-1}$) showed the highest specific growth rate (0.94 day^{-1}). This indicates that the cultures were light-limited when grown at low light intensity (31.9 $\mu Em^{-2}s^{-1}$).

Table 1. Biomass and specific growth rates of the different batches of *Ankistrodesmus convolutus*.

	Unbuffered	Buffered*		
Parameter	Low light+ Air (Control)	Low light+ Air	High Light+ Air	Low light+ 5% CO_2
Dry weight (mg/L)	122	124	150	308
Specific growth rate (day^{-1})	0.74	0.71	0.94	0.85

* + 10 mM HEPES

The unbuffered culture showed an increase of pH from 6.8 to 10.0 at the end of the study. The increase in pH could be attributed to the release of OH^- from nitrate assimilation. However, the increase in pH did not appear to have any adverse effect on the growth of *A. convolutus*. Buffering of the medium to maintain pH within 6.8 - 7.0 did not improve growth.

3.2 Biochemical Composition

The culture subjected to high light intensity afforded the highest lipid content (26.0% dry weight) (Table 2). When aerated with air alone, lipid content of the buffered culture (23.4% dry weight) was higher than the unbuffered culture (15.6% dry weight) at low light intensity.

The culture grown at low light intensity in the buffered medium and aerated with air alone had the highest protein content (23.6% dry weight). Carbohydrates constituted the lowest proportion (7.5 - 12.0% dry weight) of the algal biomass. The culture grown at low light intensity and aerated with air alone in the unbuffered medium had the lowest carbohydrate content (7.5% dry weight).

Yields of the biochemicals from the culture aerated with air containing 5% CO_2 were higher than those in the other batches due to the higher biomass attained.

3.3 Fatty Acid Composition

The predominant fatty acid of *A. convolutus* was α-linolenic acid (18:3n-3) which ranged from 73.1 - 85.7% of the total fatty acids (Table 3). Other fatty acids produced were mainly unsaturated fatty acids, namely 16:4, 18:1, 18:2 and 18:4. The only saturated fatty acids synthesised were 16:0 and 18:0.

The unbuffered culture exhibited a slightly lower proportion of 16:0 but a slightly higher proportion of 18:3n-3 compared to the other batches. The culture aerated with additional CO_2 produced less 16:0, 16:4, 18:1, 18:2 and 18:4 but more 18:3n-3 than the other cultures. No stearic acid (18:0) was produced by the CO_2-supplemented culture. Total fatty acid content remained almost constant for all the cultures (Table 3).

Table 2. Biochemical composition of the different batches of *Ankistrodesmus convolutus*.

	Unbuffered	Buffered		
Biochemical (% DW)	Low light + Air (Control)	Low light + Air	High Light + Air	Low light + 5% CO_2
Lipids	15.6 (19.0*)	23.4 (29.0)	26.0 (39.0)	17.0 (53.0)
Carbohydrates	12.0 (14.6)	7.5 (9.3)	10.0 (15.0)	12.0 (37.0)
Proteins	18.3 (22.3)	23.6 (35.4)	18.4 (27.6)	18.3 (56.4)

* Figures in parentheses represent values expressed in mg/L culture.

Table 3. Fatty acid composition of the different batches of *Ankistrodesmus convolutus*.

	Unbuffered	Buffered		
Fatty acid (% Total Fatty acids)	Low light + Air (Control)	Low light + Air	High Light + Air	Low light + 5% CO_2
16:0	7.2	5.3	5.6	4.4
16:4	1.8	2.0	2.3	0.9
18:0	0.4	0.4	1.0	-
18:1	7.7	5.1	5.4	4.0
18:2	8.2	5.7	6.5	4.0
18:3	73.1	79.6	78.6	85.7
18:4	1.6	0.8	0.8	0.5
Total fatty acids (% DW)	0.7	0.7	0.7	0.6

3.4 Pigment Composition

The pigments of this alga include lutein, violaxanthin, neoxanthin, α-carotene, ß-carotene, chlorophyll-a and chlorophyll-b (Table 4 and Figure 2). These pigments are commonly found in other green algae. However, antheraxanthin, zeaxanthin and loroxanthin which are known to be present in other green algae (11) were not detected in the present study.

A. convolutus grown at high light intensity had lower cellular content of chlorophyll-a (26.7 mg/g dry weight) than at other conditions (30.9 - 32.6 mg/g dry weight). No marked variation in the contents of other pigments was observed.

Appreciable amounts of xanthophylls (10.7 - 12.4 mg/g dry weight) with the predominance of lutein were observed. Very low contents of carotenes were detected and more α-carotene than ß-carotene was produced. The total carotenoid content (12.2 -16.0 mg/g dry weighht) of *A. convolutus* was much higher than the content reported for another species of *Ankistrodesmus* (4.1 mg/g dry weight) (12).

Table 4. Pigment composition of the different batches of *Ankistrodesmus convolutus*.

	Unbuffered	**Buffered**		
Pigment (mg/g DW)	Low light + Air (Control)	Low light + Air	High Light + Air	Low light + 5% CO_2
Chlorophyll-a	32.6	30.9	26.7	32.4
Chlorophyll-b	12.0	15.7	13.0	13.2
Chl-a:chl-b	2.7	2.0	2.1	2.5
Total chlorophylls	44.6	46.6	39.7	45.6
Lutein	5.9	5.7	4.9	5.4
Neoxanthin	2.7	2.9	2.3	3.3
Violaxanthin	3.5	3.8	3.5	3.6
Total xanthophylls	12.1	12.4	10.7	12.3
α-carotene	1.6	1.0	0.9	1.5
ß-carotene	0.9	0.7	0.6	0.9
α-car:ß-car	1.8	1.4	1.5	1.7
Total carotenes	2.5	1.7	1.5	2.4
Total carotenoids	14.6	14.1	12.2	14.7
Σcarotenoids: Σchlorophylls*	0.3	0.3	0.3	0.3

* Ratio of total carotenoids to total chlorophylls.

4. CONCLUSION

Of the four batches of cultures of *A. convolutus*, the culture aerated with air containing 5% CO_2 attained the highest biomass. Cells grown at high light intensity had the highest lipid content. The predominant fatty acid of *A. convolutus* was 18:3n-3 under all the culture conditions employed. No marked variation in fatty acid composition was shown. *A. convolutus* grown at high light intensity had a lower chlorophyll a content than grown at low intensity. No significant variation in the contents of the other pigments was observed. The predominant carotenoid of *A. convolutus* was lutein (4.9 -5.9 mg/g dry weight). Further manipulative studies to improve growth and carotenoid production are in progress.

5. ACKNOWLEDGEMENT

The above study was sponsored by a grant (R & D 1/026/01) from the Malaysian government.

6. REFERENCES

1. Cohen Z., 1986 Products from microalgae. In Richmond A. (ed), CRC Handbook of Microalgal Culture. CRC Press, Boca Raton, pp. 421-454.
2. Merchuk J.C., 1990 Tibtech. 8: 66-71.
3. Chu W.L., Phang S.M., Goh S.H. and Phang S.M., 1992 Promising microalgae for production of useful chemicals. In Shaari K., Kadir A.A., Ali A.R.M. (eds), Proc. Conf. Medicinal Products from Tropical Rain Forest Research Inst. Malaysia, Kuala Lumpur, pp. 338-345.
4. Nichols H.W., 1973 Growth media-freshwater. In Stein JR (ed), Handbook of Phycological Methods: Culture Methods and Growth Measurments. Cambridge U.P., Cambridge, 7-24.
5. Bligh E.G. and Dyer W.J., 1959 Can. J. Biochem. Physiol. 37: 911-917.
6. Christie W.W., 1989 Gas Chromatography and Lipids. The Oily Press of Scotland, pp. 69-70.
7. Bradford M.M., 1976 Anal. Biochem. 72: 248-254.
8. Kochert A.G., 1987 Carbohydrate determination by the phenol-sulphuric acid method. In Hellebust J.A. and Craigie J.S. (eds), Handbook of Phycological Methods - Physiological and Biochemical Methods. Cambridge U.P., Cambridge, pp. 95-97.
9. Fabregas J., Abalde J., Herrero C., Cabezas B.V. and Veiga M., 1984 Aquaculture 42: 207-215.
10. Goldman J.C., Bennert M.R. and Riley C.B., 1982 Biotech. Bioeng. 24: 619-631.
11. Goodwin T.W., 1980 The Biochemistry of the Carotenoids. Vol. 1. Plants. Chapman and Hall, London, pp. 207-256.
12. Paerl H.W., Lewin R.A. and Cheng L., 1984 Bot. Mar. 27: 257-264

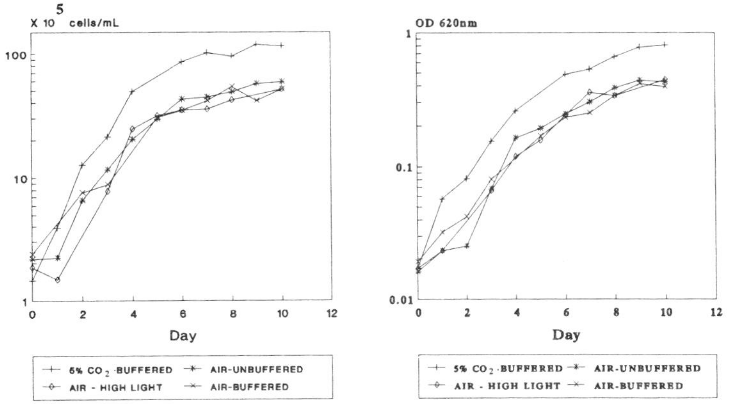

Figure 1. Growth of *Ankistrodesmus convolutus* based on a) cell density and b) OD_{620}.

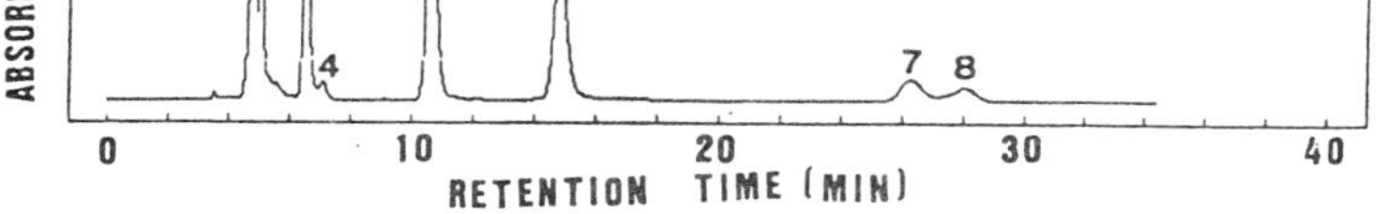

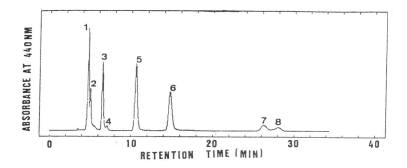

Figure 2. A typical chromatogram from HPLC analysis of pigments of *Ankistrodesmus convolutus*,
(**1**=Neoxanthin, **2**= Violaxanthin, **3**= Lutein, **4**= Unknown, **5**= Chlorophyll b, **6**= Chlorophyll a, **7**= α-carotene, **8**= **ß-carotene)**

HIGH PRODUCTIVITY OF THERMOSTABLE XYLANASE FREE OF CELLULASE: A PROMISING SYSTEM FOR LARGE SCALE PRODUCTION

M. M. Hoq*, C. Hempel and W.- D. Deckwer
GBF - Gesellschaft für Biotechnologische Forschung mbH, Biochemical Engineering Division, Mascheroder Weg 1, D-38124 Braunschweig, GERMANY

Three thermophilic fungi, *Thermomyces lanuginosus* RT9, *Thermomyces lanuginosus* MH4 and *Rhizomucor pusillus* MH3 were examined for the production of xylanase free of cellulase in xylan or lignocellulosic-based media. The *Thermomyces* strains produced cellulase-free xylanase while *R. pusillus* had traces of cellulase in addition. *T. lanuginosus* RT9 displayed the highest xylanase productivity under nearly all cultivation conditions. Partial optimization of cultivation conditions with *T. lanuginosus* RT9 produced an increase of xylanase activity from 3156 nkat/ml to 26,170 nkat/ml in shake flask culture. In the bioreactor system, the lag phase reduced considerably over shake culture system and resulted in at least 5-fold higher productivity (45,703 U/l/h). The xylanase worked best at 70°C and pH between 6.5 to 7.0. The activity was almost constant at 55°C for 67 h and at room temperature for at least 3 months and in the pH range 5-9.

1. INTRODUCTION

In recent years, growing interest has been generated in xylanases free of cellulases in order to remove xylan selectively from cellulose preserving cellulose fibre length. These applications include its usages in pulp and paper industry mainly in the production of dissolving pulp (Paice et al (1)) and pretreatment of pulp for boosting bleaching process (Paice et al (2)) which reduces or replaces the usage of chlorine which is ecologically harmful. Xylanase free of cellulase is particularly useful for the selective removal of xylan (a hardening component) from low quality jute fibre enabling them to spinning for the effective utilization of the total jute production (Hoq et al (3)).

Xylanases of high thermostability are of particularly suited for such applications. Therefore, in this studies, attempt has been made to develop a process for production of cellulase-free thermostable xylanase and its recovery, purification and technical applications. The present report describes the efficiency of xylanase production by three thermophilic fungi and focuses on the production and some properties of a very high level of xylanase by *T. lanuginosus* RT9.

2. EXPERIMENTAL

2.1 Microorganisms

The *Thermomyces lanuginosus* RT9, *Thermomyces lanuginosus* MH4 and *Rhizomucor pusillus* MH4 isolated in Bangladesh and identified from Deutsche Sammlung von Mikroorganismen (DSM), Braunschweig, Germany, were used in this experiment. All these organisms were grown on Potato dextrose agar slants at 50°C for about 5 days and stored at room temperature, with subculturing every 4 or 5 weeks.

---

*Permanent address: Department of Microbiology, University of Dhaka, Dhaka 1000, Bangladesh

2.2 Media and shake culture conditions

Three known media for cultivation of xylanolytic fungi, were used with some modifications. Wheat bran, corn cobs or xylan was used as carbon sources. These were
Medium 1. Basic mineral medium (without $CaCl_2.2H_20$) of Mandels (4) containing 2% wheat bran, or 1.0% xylan, 1.0% peptone, 0.2% $(NH_4)_2S0_4$, .03% $MgSO_4.7H_20$, 0.1% tween-80, and 1.5% KH_2P0_4 in distilled water.

The following Medium 2 and Medium 3 developed by Gomes et al. (5,6) for the cultivation of *T. lanuginosus,* were used in this study with some modifications. **Medium 2.** 1.5% yeast extract, 0.21% $(NH_4)_2S0_4$ 0.03% $MgS0_4$ 0.03% $CaCl_2.2H_20$, 0.05% $FeS0_4.7H_20$ an 1.0% KH_2P0_4 in tap water. **Medium 3**. 0.6% meat peptone, 0.12% $(NH_4)_2S0_4$ and 1.5% KH_2P0_4 in tap water.

Each culture medium of 100 ml in 300 ml baffled -Erlenmeyer flask was inoculated with actively growing 4-5 days old colony of the organism. For each culture condition, at least 3 replica flasks were made. The flasks were than shaken at 55°C or 50°C on an orbital shaker at 120 or 200 rpm for 5 to 7 days unless otherwise stated. Samples were taken at regular interval of time. The culture filtrates were centrifuged and the clear supernatant was used for assays of enzymes and soluble protein.

2.3 Enzyme assays

Xylanase activity was determined according to Bailey et al. (7) using 1% birch wood xylan (Sigma, lot: 52H0544 X0502) in 0.05 M Na-citrate buffer (pH 6.0 or 6.5) after 5 min reaction time. The assay mixture containing 1.8 ml substrate and 0.2 ml suitably diluted enzyme solution in the buffer was incubated at 50°C for 5 min and the reaction was stopped by addition of 3 ml DNS reagent followed by keeping in boiling water for 15 min.The amount of reducing sugar liberated was determined by DNS method using xylose (Sigma) as standard. One unit of xylanase activity was defined as 1 nmol of xylose equivalents released per second (nkat). Carboxy methyl cellulase (CMCase) was assayed using carboxy methyl cellulose(Sigma) as described by IUPAC (8). ß-xylosidase were assayed by the method of Herr et al. (9) using p-nitrophenyl-ß-D-xyloside (Sigma). One unit of ß-xylosidase activity was defined as the amount of enzyme which catalyze the release of 1 nmol p-nitrophenol per sec.

2.4 Other analyses

Soluble protein in the culture filtrate was analyzed by the method of Bradford (10). Reducing sugar in the assay mixture and free sugar in the medium were estimated by the method of Miller (11). Total dry matter (TDM) was determined by centrifuging 5 or 10 ml sample in a preweighed falcon tube. The retenate was washed once with distilled water. After recentrifuging, the retenate was dried under vacuum at 40°C for 48h and reweighed.

3. RESULTS AND DISCUSSION

Xylanase production by the thermophilic fungi under shake culture : To obtain a potential producer of xylanase, the three fungi were cultivated in Mendel's medium (Medium 1) on wheat bran or xylan as carbon source at 55°C and pH 6.0. The xylanase productivity is shown *Figure 1*. Maximum xylanase productivity (998 U/l/h) was demonstrated by *R. pusillus* followed by *T. lanuginosus* RT9 (773 U/l/h) on wheat bran while in xylan-induced medium its productivity was

Table 1. Production of xylanases by *T. lanuginosus* strains on different substrates under shake culture*.

Substrates (W/V)	Xylanase activity (nkat/ml)	
	T. lanuginosus RT9	*T. lanuginosus MH4*
Xylan (2%)	24600	6378
Corn cobs (3%)	9641	2330
Wheat bran (3%)	3177	1246

* Results of 7 days cultivation at 50°C, pH 6.5, shaking at 200 rpm

highest with *T. lanuginosus* RT9 . T. lanuginosus MH4 showed much lower xylanase production compared with other cultures. No cellulase activity was detected with either of the two *T. lanuginosus* strains in all cultivation conditions employed. This results conforms our previous finding with T. lanuginosus RT9 under solid state culture (Hoq et al (3)) and the result of Gomes et al. (5). *R. pusillus* produced a little amount of cellulase.

From this initial study, only the *T. lanuginosus* strains being the producers of xylanase free of cellulase, were compared for the xylanase production in the medium containing yeast extract (Medium 2) instead of peptone on 3% each of wheat bran or corn cobs with a view to examine the efficacy of cheap lignocellulosic material as substrate. Both strains grew well in this medium but displayed much differences in xylanase production under the conditions employed. The reason may be that the cultivation conditions with the *T. lanuginosus* MH4 was less favourable. *T. lanuginosus* RT9 showed higher level of xylanases on all substrates as shown in *Table 1*. The corn cobs which were available for our use were from edible and immatured corn. The matured corn cobs induced comparable xylanase production to xylan as reported by Gomes et al. (5). Finally *T. lanuginosus* RT9 turned out to be the best xylanase producer and was selected for further studies.

The production of xylanase in shake culture was carried out in a partially optimized medium 2 on 2% xylan, at temperature 50°C and pH 6.5. Higher level of xylanase production was achieved after 7 days (*Figure 2*). This result attained a 3.5 times higher productivity than one with Mendel's medium which was at higher temperature and lower pH.

Xylanase production in bioreactor: The same shake culture conditions were used in 15 l bioreactor (STR) under varied aeration and agitation rates with a view to regulate optimum oxygen concentration level, maximizing enzyme production. The xylanase production was highly influenced by the dissolved oxygen concentrations in the culture (related results are not included). A typical result is as displayed in *Figure 3*. After 7 h, the production of xylanase started and simultaneously, the dissolved oxygen concentration (pO_2) reduced to about 8% which continued till about 20 h. This prolonged reduced pO_2 level was due to active biomass growth as indicated from the course of soluble protein production. The course of xylanase production and its maximum level (18, 318 nkat/ml) after 24 h suggest that the enzyme production was growth associated. The lag phase was always much shorter than the one with shake culture system thus resulting in 5-fold higher productivity (43,703 U/l /h). This is the highest xylanase productivity on xylan so far reported. In this connection, a comparison on the productivity of xylanases by some of the best processes as carried out in bioreactor are shown in *Table 2*.

Table 2 Comparison of volumetric productivity of xylanases by different bioprocesses

Microorganism	Productivity (U/L/h)	References
Aureobasidum pullulans CBS 58475	300	12
Trichoderma reesei QM 9414	1,180	13
Schizophyllum commune	23,916	14
T. lanuginosus DSM 5826	11,180	6
T. lanuginosus RT9	42,000	Present work

Properties of xylanase: The temperature optimum and thermal stability of the xylanase are shown in *Figure 4*. For temperature optimum, the enzyme activity was determined by incubating with 1% xylan for 5 min in citrate buffer (pH 6.5) at different temperatures. The optimum temperature for the enzyme activity was 70°C, indicating its thermophilic nature. The thermostability was tested by heating the enzymes for different times and temperatures and the residual activity was determined at 50°C. The enzyme activity was constant at 55°C for at least 67 h and under prolonged storage at room temperature (22 to 25°C) for 3 months. The pH profile and pH stability are shown in Figure 5. The optimum pH of the enzyme for its activity displayed between pH 6.5 to 7.0. The pH stability of the enzyme was examined by keeping its buffer solutions at 4°C for 24 h and assayed at 50°C. The enzyme displayed a wide pH stability between 5 to 9.

4. CONCLUSION

The highest productivity of cellulase-free xylanase was achieved by a wild strain of *Thermomyces lanuginosus* RT9. The characteristics of wide thermo- and pH stabilities of the enzyme are unique for its storage and technical applications.

5. ACKNOWLEDGEMENT

M. M. Hoq is thankful to Alexander von Humboldt Research Foundation for the award of a fellowship to enable him to carry out this research.

6. REFERENCES

1. Paice, M.G., Bernier, R. and Jurasek, L. (1984) *J. Wood Chem. Technol.* 4, 187-198

2. Paice, M.G., Bernier, R. and Jurasek, L. (1988) *Biotech. Bioeng.* 32, 235-239

3. Hoq, M. M., Alam, M. Mohiuddin, G. and Gomes, I. (1992) In *Harnenessing Biotechnology for the 21st Century,* ACS, Michael , L. and Bose, A. (eds), pp. 568-573

4. Mandel, M. and Andreotti, R. and Roche, C. (1976) *Biotech. Bioeng.* 6, 21-33

5. Gomes, J. Purkarthofer, Hyn, M., Kapplmüller, J., Sinner, M. and Steiner, W. (1993) *Appl. Microbiol. Biotechnol.* 39, 700-707

6. Gomes, J., Gomes, I., Kreiner, W., Esterbauer, H., Sinner, M. and Steiner, W. (1993) *J. Biotechnol.* 283 - 297.

7. Bailey, M.J., Biely, P., and Poutenan, K. (1992). *J. Biotechnol.* 23:257-270.

8. IUPAC (International Union of Pure and Applied Chemistry) (1987) Measurement of cellulase activities. *Pure Appl. Chem.* 59:257-268.

9. Herr, D., Baumer, F., Dellweg, H. (1978) *Appl. Microbiol. Biotechnol.* 5:29-36.9.9.

10. Bradford, M. M. (1976) *Anal. Biochem.* 72:248-254.

11. Miller, G. L., *Anal. Chem.* 31:426-428.

12. Priem, B., Dobberstein, J. and Emois, C. C. (1991) *Biotechnology lett.* 13(3), 149-154

13. Poutannen, K. Rättö, M., Puls, J. and Viikari, L. (1987) *J. Biotechnol.* 6, 49-60

14. Haltrich, D., Preiss, M. and Steiner, W. (1993) *Enzyme Microbiol. Technol.* 15, 854-860

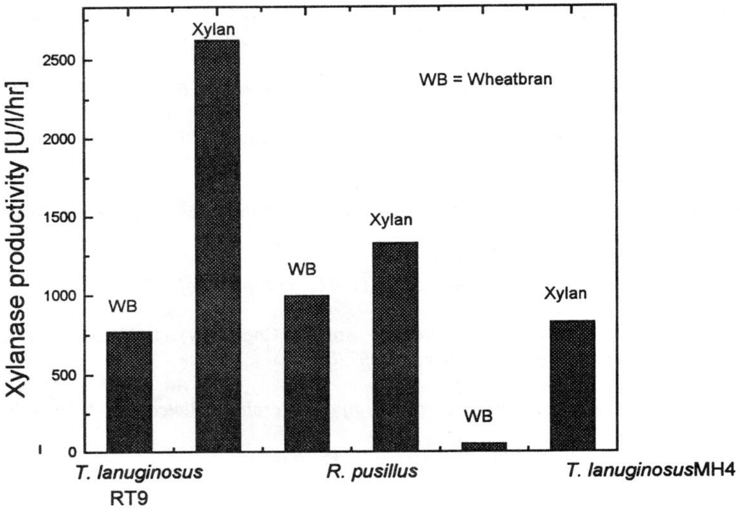

Fig. 1 Xylanase productivity by three fungi in Mendel's medium on wheat bran or xylan. Cultivation at 55°C, pH 6.0 and shaking at 120 rpm.

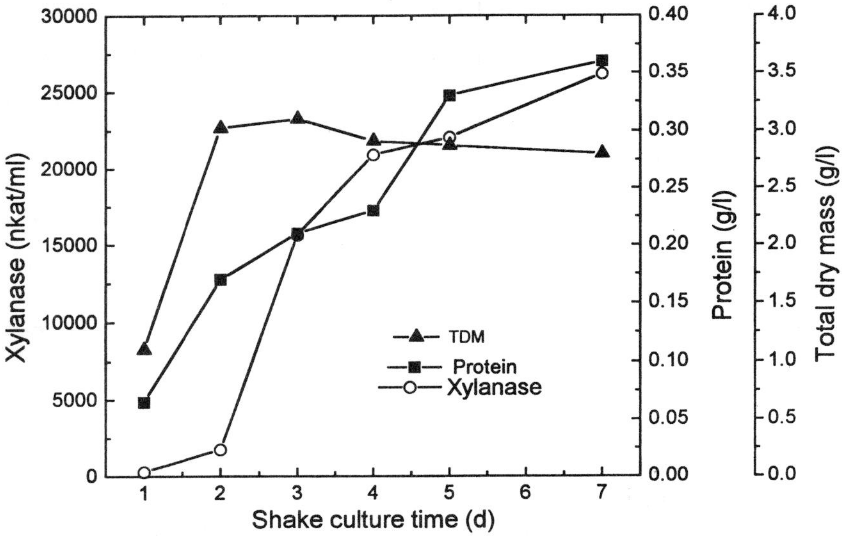

Fig. 2 Time courses for xylanase, biomass and extracellular protein production by *T. lanuginosus* RT9 on xylan. Cultivation at 50°C, pH 6.5 and shaking at 200 rpm.

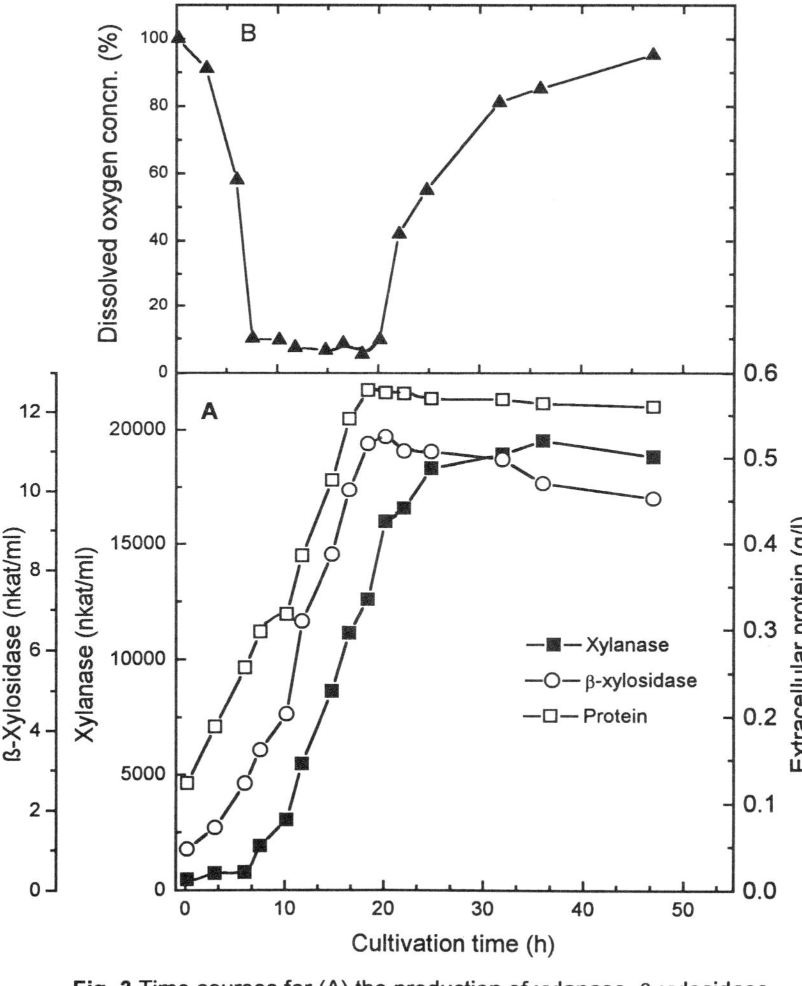

Fig. 3 Time courses for (A) the production of xylanase, β-xylosidase and extracellular protein and (B) changes in dissolved oxygen concentration in 15 l bioreactor by *T. lanuginosus* RT9 on xylan. Cultivation at 50°C, pH 6.5, aeration 0.5 vvm and agitation at 200 rpm

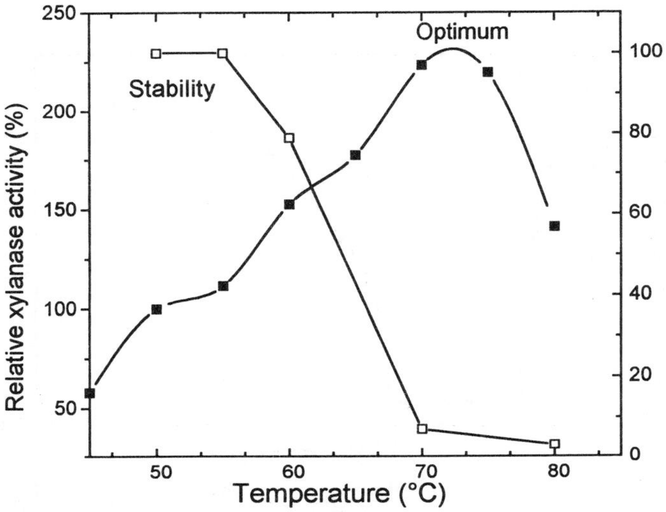

Fig. 4 Temperature optimum and thermostability of *T. lanuginosus* RT9 xylanase

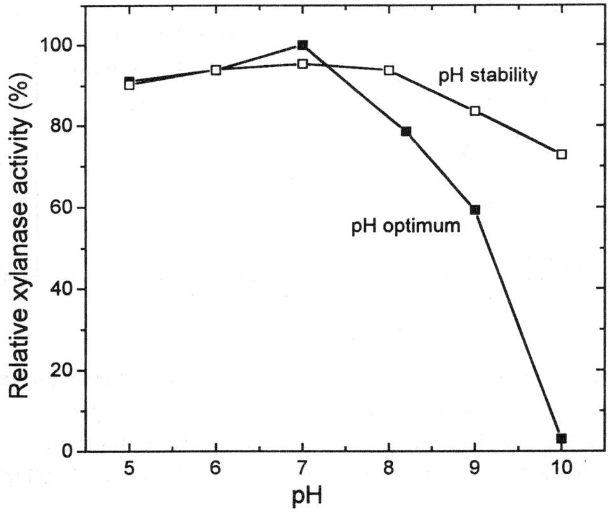

Fig. 5 pH optimum and stability of *T. lanuginosus* RT9 xylanase

MODELLING AND OPTIMISATION OF CELL CULTURES.

Charles Sanderson, John Barford and Geoff Barton.
Department of Chemical Engineering, University of Sydney, NSW 2006 Australia.

A number of potential applications for a detailed computer simulation are highlighted, and a model capable of addressing them is described. This model, which has been "tuned" to two different cell lines, is found to match the experimental results very closely. Two simple feeding strategies are applied to these simulated batch cultures and it is found that the predicted profitability of antibody production maay be increased by up to 250%.

1. INTRODUCTION.

Over the past ten years a dynamic model for cells growing in culture has been developed. It has been applied primarily to animal, hybridoma cells and calculates the rate of antibody production, the cellular growth rate and the rate of formation/utilisation of metabolites Barford *et al* (1). The main aim of the project has been to provide a unified basis for the various disparate observations of cellular metabolism: by combining this knowledge into a computer model, a powerful tool has been developed which has a number of potential applications:-

Simulation: A good model is relatively cheap, simple and quick to run. The results are self consistent and are far easier to produce than comparable experimental measurements.

Data Analysis: Since the model predictions are self consistent, they may be compared to experimental results in order to highlight errors in the experimental data.

Theory Testing: The model includes all the main aspects of cellular metabolism and is modular in form, so new theories for metabolic phenomena can easily be incorporated. This both verifies that the theory has the expected effects and improves the simulation.

"What if ?": The model can be used to predict the effects of changing the cell's metabolism or culture conditions. An example, that has been tried by this group, Johnson *et al* (4), is to examine the effect of viral infection on a cell's growth and antibody production rates.

Soft Sensing: The model can be combined with readily available measurements, such as medium concentrations, to allow difficult-to-measure variables, eg internal reaction rates, to be inferred.

Advanced Control: A model allows various forms of model based control to be implemented on a bioreactor. These are discussed in more detail by Fu and Barford (3).

Optimisation: The model can be used to predict better medium formulations and feeding policies. A simple study into possible feeding policies is described here - more rigorous optimisation could be achieved by linking the model to a numerical optimisation package.

2. MODEL STRUCTURE.

The simulation has been written in SpeedUp, a workstation based flowsheeting package which contains excellent differential/algebraic equation solution algorithms, combined with a simple user interface and extensive error checking facilities. It also includes optimisation and parameter estimation algorithms and can easily be linked to other software.

Figure 1 gives an overview of the metabolic pathways included in the model. For simplicity, only the key metabolites have been included - these are defined as those which are either transported directly across a cellular membrane or which occur immediately before forks in metabolic pathways.

Glycolysis, the pentose phosphate pathway, glutaminolysis and the tricarboxylic acid cycle have all been included in some detail. They provide the cell with not only energy, but also a means of synthesising biosynthetic "precursors", such as amino acids or phospho-ribosyl pyrophosphate (PPP). These precursors are combined to form new cells, viral particles or antibody. Michaelis-Menton type kinetics describe the rates of reaction and of transport across cellular membranes.

Unlike previous versions of this model, Barford *et al* (1), the amino acids are treated independently and the concentrations of all the metabolites are tracked, allowing a more advanced simulation to be developed. The independent treatment of amino acids means that their transport across the cellular boundary can be modelled with the transport groups suggested by Christensen (2). The amino acids are sub-divided into eight groups (see Figure 1) with the members of each group competing for transport across the cell wall - no competition is assumed to occur with amino acids from other groups. The other metabolites are assumed to cross the cell wall independently, while transport across the mitochondrial membrane occurs by both independent and antiport means.

Tracking internal pool size means that feedback inhibition may be included, making the model more realistic by ensuring that metabolism remains in balance. As energy ceases to limit growth, for example, the intracellular ATP concentration rises and inhibits the energy producing reactions. Knowledge of the concentration of the metabolite immediately before a split in a pathway also allows the extent to which the possible branches are utilised to be varied, based on the Michaelis constants (k_m) for the competing reactions; the one with the lower k_m will be used preferentially until its maximum rate of reaction (V_m) is approached. The second pathway is then used more extensively. In this model, for example, pyruvate will enter the mitochondria more readily than form lactate - lactate formation may be viewed as a sink for excess pyruvate.

It should be noted that the model is quite large, consisting of some five hundred algebraic and one hundred differential equations. However, the model fits comfortably into the memory of a small workstation and can be solved in around ten minutes. The model's complexity means that it can be used in the wide range applications described previously.

There are also some three hundred user defined parameters. With such a large number of "free" variables, there is a danger that the model will be over-tuned and so be fitted to the set of training data rather than being a true representation of the modelled system. In fact, relatively few of the variables are truly free; all of the variables are bounded, several of the parameters are directly measurable (such as the composition of cellular material) and the effect of many more parameters on the model's prediction is only weak.

One major advantage of the modular approach employed in the simulation is that, as information becomes available, the model can be made increasingly more realistic. Model extensions currently under consideration include more detailed treatment of new cell and product formation, verification of cellular composition and inclusion of cell cycle effects.

3. MATCHING MODEL PREDICTIONS TO EXPERIMENTAL RESULTS.

There are two detailed sets of data against which this model has been tested, one produced at the University of Sydney by Marquis (6) for the AFP27, murine cell line and the other by Kian Fong at the National University of Singapore (5) for the murine, 2HG11 line. Figures 2 to 5 show the measured and predicted biomass, antibody, glucose, lactate, alanine, glutamine and ammonia concentration in the medium, for the AFP27 line. Figures 6 to 9 show similar plots for the 2HG11 line. It can be seen, in both cases, that the tuned model and the experimental data agree very well. Similar agreement is observed for the other measured amino acid profiles (not shown).

It can be seen that the AFP27 line produces more alanine and uses glutamine more slowly than 2HG11. It also grows more quickly and produces antibody at a higher specific rate. In both cases, the model fails to produce lactate and consume glucose quickly enough initially, though the final values are in close agreement with the experimental data. The near linear changes in these profiles during the growth phase are extremely difficult to reconcile with the other measured variables using the current model structure, an indication that either some unmodelled effect is occurring, or that the experimental data is inaccurate. Glutamine depletion causes cell death to begin for both cell lines, while for the 2HG11 cell line, most of the glucose was also consumed.

The predicted ammonia concentration for the 2HG11 cell line is higher than that observed experimentally, while the reverse is true for AFP27. The modelled ammonia concentration depends on a number of factors, including the cellular composition and the rates of glutaminolysis and alanine production, and so it is one of the more susceptible predictions to modelling error. These profiles also highlight the problems caused by experimental error: both the initial estimates and the variability of the data are important during training. If the initial point for the AFP27 ammonia profile were higher, the prediction would match the measurements far better, while the fairly noisy 2HG11 experimental data calls its accuracy into question .

There is also the suggestion in the experimental data that dead cells may break up: this effect has not, as yet, been included in the model structure.

4. FEEDING POLICIES.

The simulation was used to investigate the effects of two feeding policies on the two cell lines for which the model had been tuned. As the first policy, the same initial medium as for the batch cultures was used and to that, 50mM glucose and glutamine solutions were independently added at fixed rates, a moderately simple control policy to implement. In the second policy, the medium concentrations of glucose and glutamine were held constant by continuously adding the required amount of 50mM glucose or glutamine solution; the initial conditions were otherwise identical to those of the batch case. This represents a more complex, and therefore expensive to implement, control scenario. The initial batch volume was ten litres in all cases.

The results of this investigation were analysed by comparing the maximum antibody concentration and the maximum profitability, subject to the constraint that the medium volume was not allowed to more than double. The results are displayed in Figures 10 to 17 as three dimensional plots with contours of the surfaces projected onto the base. Table 1 details the values of the contours used.

TABLE 1. Contours on Three Dimensional Plots.

Figure	Shows	Units	Contours
10, 14	Profitability	\$/h	300, 285, 255, 215
11, 15	[Ab]	μM	0.550, 0.525, 0.500, 0.450, 0.300
12, 16	Profitability	\$/h	330, 320, 310, 300, 275
13, 17	[Ab]	μM	0.740, 0.675, 0.600, 0.520

The profit function, described below, is intended to provide a simple comparison for the value of the culture broth. It does not include fixed or downstream purification costs and is dominated by the amount of antibody produced and the batch time.

$$Profitability\ (\$/h) = \frac{V_{ab}*m_{ab} - V_{inoc} - V_{med} - V_{gl}*q_{gl} - V_{gln}*q_{gln}}{t_b + t_d}$$

where m_{ab} mass of antibody formed (mg) — V_{ab} value per mass of antibody (\$/mg)
q_{gl} volume of glucose solution added (l) — V_{inoc} value of initial inoculum (\$)
q_{gln} volume of glutamine solution added (l) — V_{med} value of initial medium (\$)
t_b batch time (h) — V_{gl} value of glucose feed (\$/l)
t_d down time between batch runs (h) — V_{gln} value of glutamine feed (\$/l)

Figures 10 and 11 show the effects on profitability and antibody concentration of feeding glucose and glutamine solution to the AFP27 culture at a fixed rate. It can be seen that glucose solution acts only as a diluent, reducing the antibody concentration and, at higher feed rates, reducing the profitability, due to violation of the volume constraint. The high initial glucose concentration in the batch means that other nutrients run out long before glucose could limit growth or product formation. Glutamine addition has a far greater effect; at low feed rates ($3x10^{-4}$ l/h) the antibody concentration reaches a maximum of 1.1μM, 3.5 times that of the batch culture; there is a corresponding local maximum in profitability of \$252/h (almost twice that of the batch culture). At slightly higher feed rates, more biomass is formed and this, combined with the diluting effect of additional feed, causes the antibody concentration to drop rapidly. At higher feed rates still, the antibody formation rate increases, causing the profitability to rise again, though its medium concentration remains low. This effect is due to the dilution of inhibitory metabolic products, such as ammonia, allowing biomass and antibodies to be produced faster. Eventually, the broth volume constraint is reached, fixing the maximum profitability at \$303/h with a glutamine feed rate of 0.09 l/h. At higher feed rates, the culture volume constraint becomes active before the batch time when significant production of antibody has ceased, so the profitability declines rapidly.

Figures 12 and 13 show the results for the scenario with constant medium concentrations of glucose and glutamine. It can be seen that the maximum profitability (\$335/h) is achieved with a relatively low glucose concentration (600μM) and a high glutamine concentration (14000μM). This larger profitability, around 10% higher than the optimum in the fixed feed rate strategy, suggests that the more difficult control scheme may be worth implementing. The maximum antibody concentration is twice that achieved by batch culture and occurs with glucose and glutamine concentrations of 600μM and 11000μM, respectively. It is interesting to note that, in this case, profitability is unconstrained - the optimum does not occur at the volume limit.

Figures 14 and 15 show the constrained profit function and antibody concentrations for the 2HG11 cell line under the fixed feed rate policy. Like AFP27, there is a "ridge" of local maximum profitability, about 50% higher than the batch culture value of \$143/h, corresponding to low feed rates of glutamine solution. Also like AFP27, the global maximum, of \$241/h, occurs at the maximum culture volume constraint, with glucose and glutamine feed rates of 0.025 and 0.06 l/h respectively. It can be seen that the highest antibody concentration, of 790μM, occurs close to batch culture conditions, when the feed rates of both glucose and glutamine are relatively small.

Figures 16 and 17 show the effect of holding the medium concentrations of glucose and glutamine constant on 2HG11. The maximum profitability occurs at a glucose and glutamine concentration of 25000μM and 7500μM, respectively. The maximum of \$244/h, as compared to the \$241/h for the fixed feed rate policy, suggests that the more advanced control is not worthwhile in this case. The maximum antibody concentration is found with the glucose concentration at less than 5000μM and glutamine at around 12500μM.

Both glucose and glutamine were almost depleted under batch culture conditions for 2HG11, so feeding either or both of these nutrients allows more antibody to be formed, though the increased culture volume causes it to be diluted.

5. CONCLUSIONS.

A number of possible applications for a detailed cell culture simulation have been identified and a simulation has been tuned to two different cell lines. Its predictions were found to compare favourably with the available experimental measurements and an initial investigation into the feeding of batch cultures for both cell lines has indicated that substantial increases in profitability (up to 250%) are achievable.

A more realistic profit function, which included the costs of purification and control, would undoubtedly affect these results. The examination of a wide range of scenarios is, however, straight forward, given the availability of a tuned model within the SpeedUp environment.

6. ACKNOWLEDGMENTS.

The partial funding of this work by the Australian Research Council is appreciated.

7. REFERENCES.

1. Barford, JP, Phillips, PJ and Harbour, C , 1992, Cytotechnology, 10, pp 63-74
2. Christensen, NH, 1989, Methods in Enzymology, 173, pp 576-616.
3. Fu, PC and Barford, JP, Cytotechnology, (In Press)
4. Johnson, L, Phillips, PJ, Harbour, C and Barford, JP, 1993, Proc. JAACT '93, Nogoya
5. Kian Fong, 1992, PhD thesis, National University of Singapore
6. Marquis, CP, 1993, PhD thesis, University of Sydney

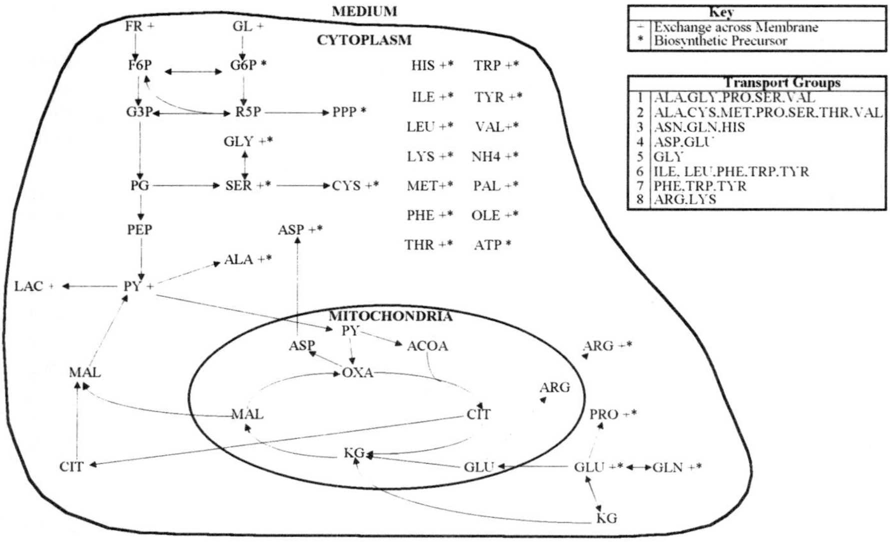

Figure 1. Overview of Modelled Internal Pathways

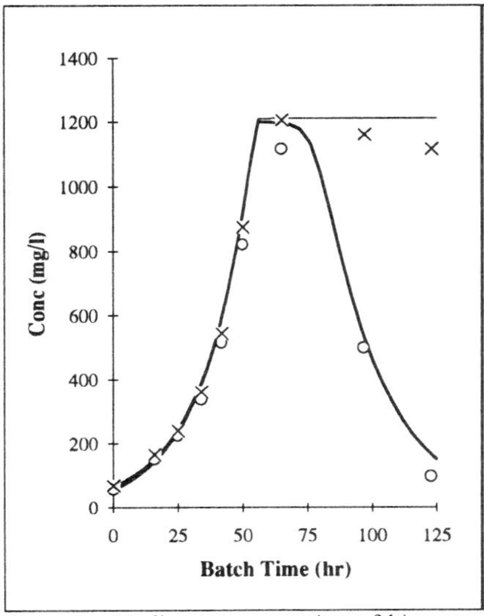

Figure 2. Medium concentration of biomass for AFP27 cell line. o - viable cells, x - total cells, solid lines - model predictions.

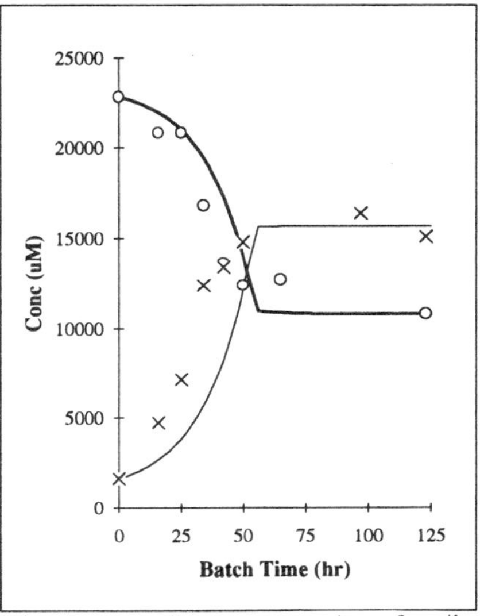

Figure 3. Medium concentration of antibody (o) for AFP27 cell line. Continuous line - model prediction

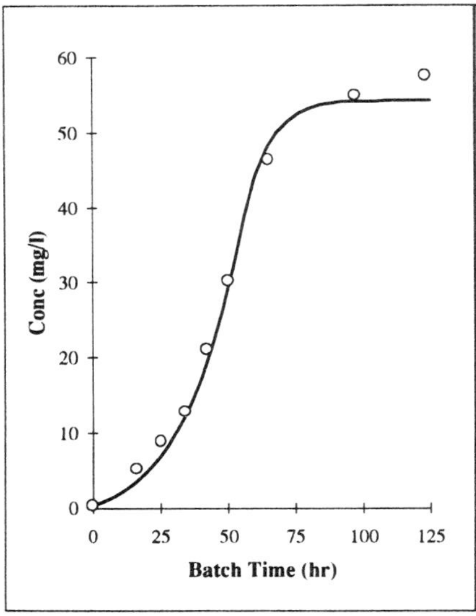

Figure 4. Medium concentration of glucose (o) and lactate (x) for AFP27 cell line. Continuous lines - model predictions.

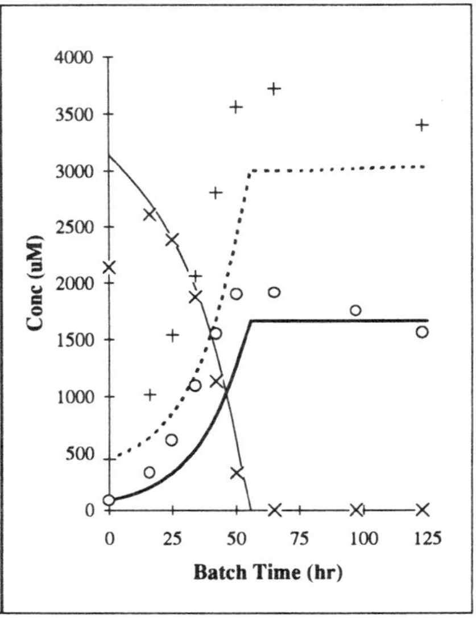

Figure 5. Medium concentration of alanine (o), glutamine (x) and Ammonia (+) for AFP27 cell line. Continuous lines - model predictions

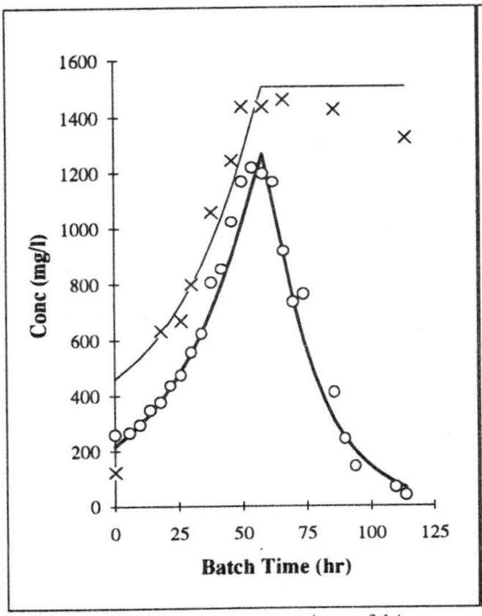

Figure 6. Medium concentration of biomass for 2HG11 cell line. o - viable cells, x - total cells, solid lines - model predictions.

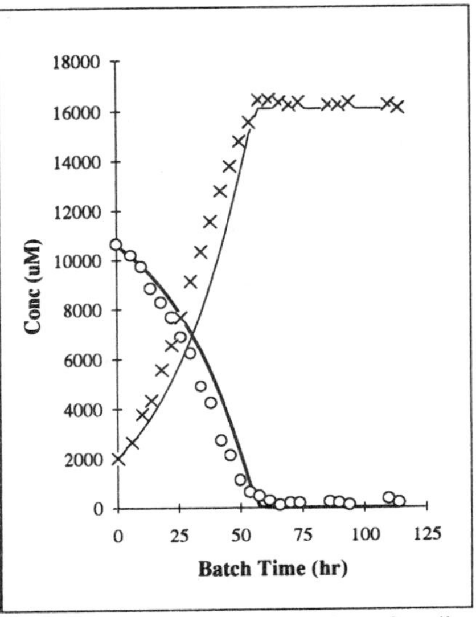

Figure 7. Medium concentration of antibody (o) for 2HG11 cell line. Continuous line - model prediction

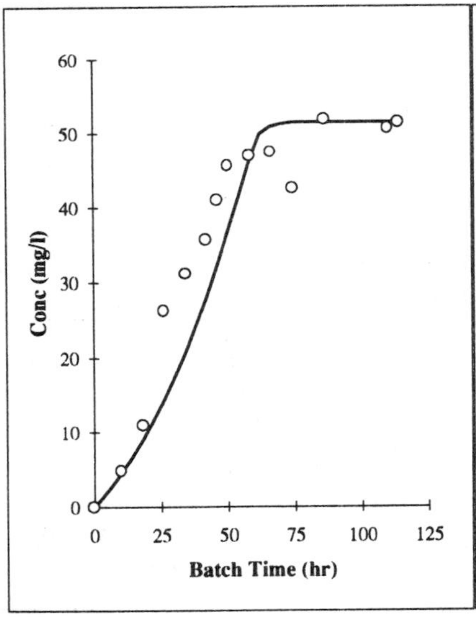

Figure 8. Medium concentration of glucose (o) and lactate (x) for 2HG11 cell line. Continuous lines - model predictions.

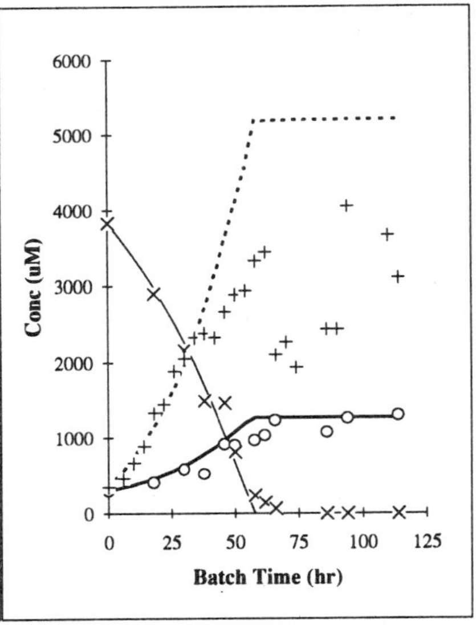

Figure 9. Medium concentration of alanine (o), glutamine (x) and ammonia (+) for 2HG11 cell line. Continuous lines - model predictions

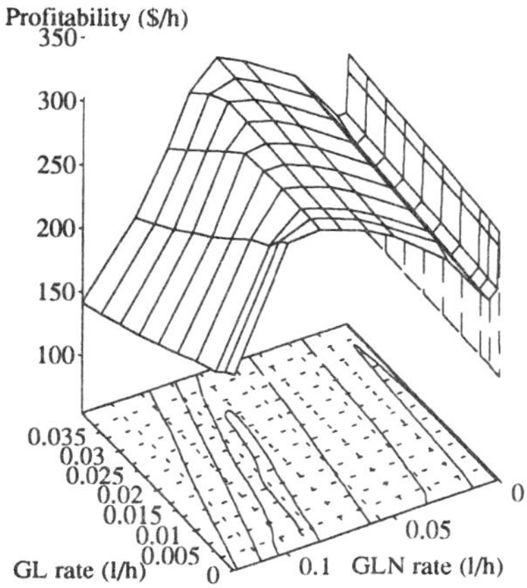

Figure 10. Profitability predicted for AFP27 cell line if batch supplemented glucose and glutamine solutions at constant feed rates

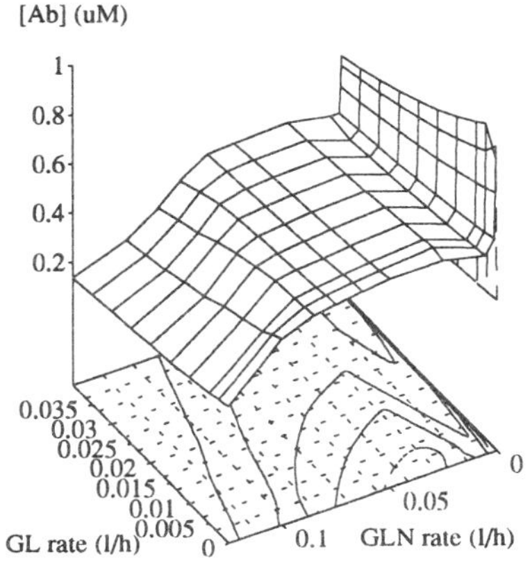

Figure 11. Antibody concentration predicted for AFP27 if batch supplemented glucose and glutamine solutions at constant feed rates

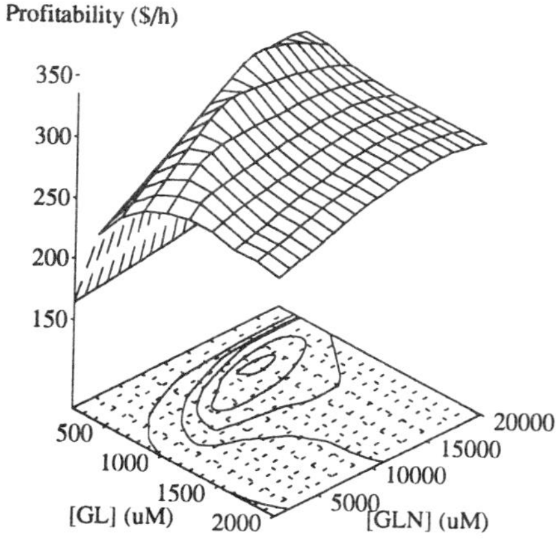

Figure 12. Profitability predicted for AFP27 cell line if glucose and glutamine medium concentrations held constant during culture.

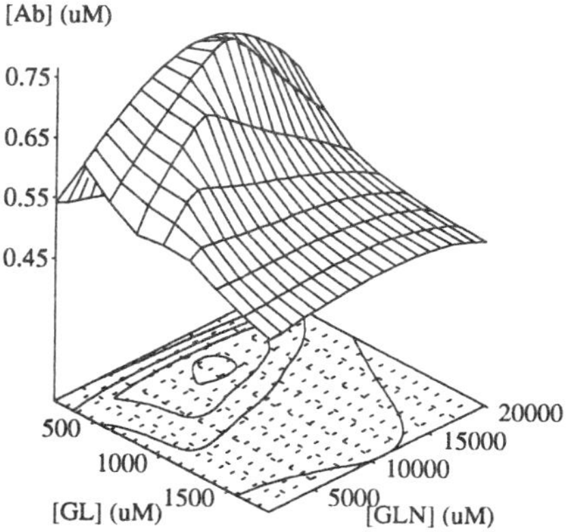

Figure 13. Antibody concentration predicted for AFP27 cell line if glucose and glutamine concentrations held constant during culture.

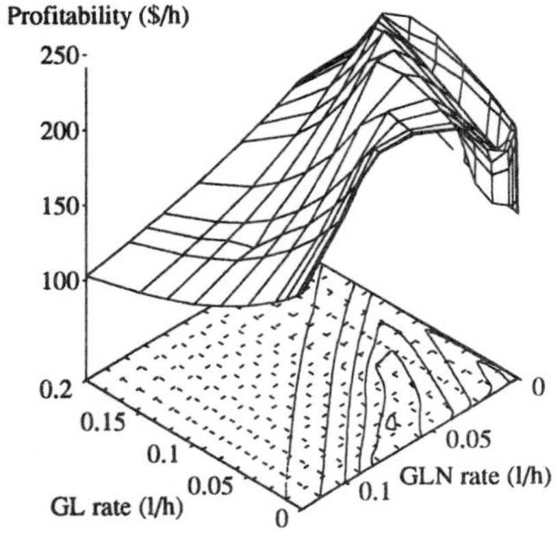

Figure 14. Profitability predicted for 2HG11 cell line if batch supplemented glucose and glutamine solutions at constant feed rates

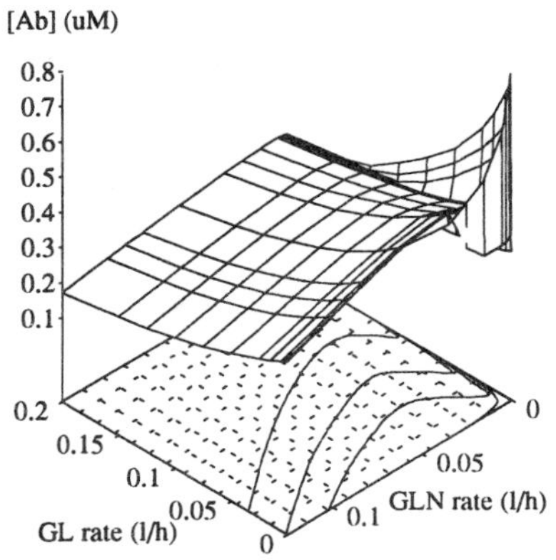

Figure 15. Antibody concentration for 2HG11 cell line if batch supplemented glucose and glutamine solutions at constant feed rates

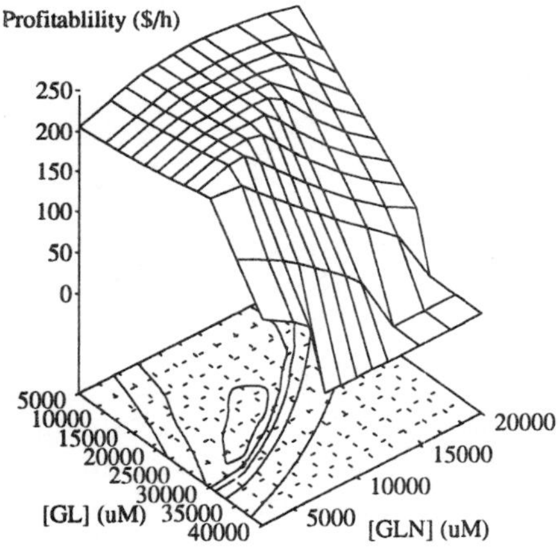

Figure 16. Profitability predicted for 2HG11 cell line if glucose and glutamine medium concentrations held constant during culture.

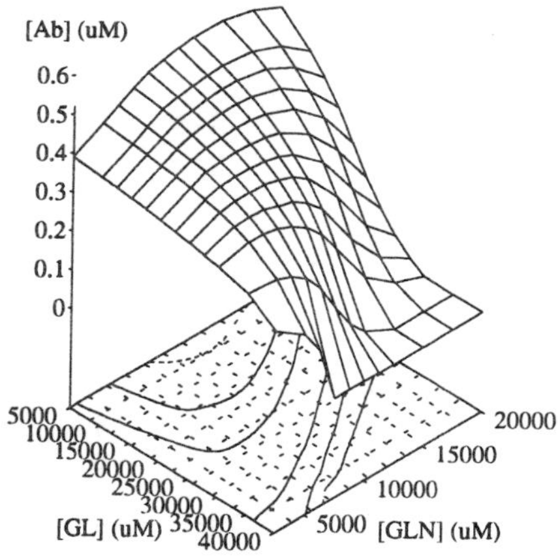

Figure 17. Antibody concentration predicted for 2HG11 cell line if glucose and glutamine concentrations held constant during culture.

PROTEIN ENRICHMENT OF CORN STOVER USING *NEUROSPORA SITOPHILA*

U. C. Banerjee, Y. Chisti and M. Moo-Young
Department of Chemical Engineering, University of Waterloo, Waterloo, Ontario, Canada N2L 3G1

The effects of solid-substrate characteristics (particle size, pretreatment conditions) on microbial biomass protein production and cellulose utilization by *Neurospora sitophila* (ATCC 36935) were investigated. Corn stover ground to various particle size fractions was the test substrate. The pretreatment utilized sodium hydroxide for delignification and hemicellulose removal. Cellulose utilization by the fungus and the crude protein production increased with decreasing substrate particle size and with increasing sodium hydroxide concentration in the pretreatment step. Under the best conditions, *ca.* 90% of the initial cellulose was consumed by the fungus and the crude protein concentration in the dry product exceeded 50% by weight.

1. INTRODUCTION

Protein enrichment of agricultural lignocellulosic residues (*e.g.*, straw, corn stover, sugarcane bagasse) for food or animal feed is potentially useful in reducing the environmental impact of these residues and in enhancing animal and human food supplies. At present under-utilized, these residues can be upgraded to food by improvements in digestibility, nutritive value and palatability by fermentation with cellulolytic microorganisms (1). Although many cellulose-degrading microorganisms, mostly fungi, are known, few would qualify as food- or feed-grade. Thus, *Penicillium funiculosum* (2), *Alternaria alternata* (3), *Trichoderma viride* (4) and *Chaetomium cellulolyticum* (5, 6) are all inappropriate for protein enrichment processes; however, the microfungus, *Neurospora sitophila*, which has a long history of use as food in oriental preparations such as *ontjom* (7 - 9), is particularly suited for such a process. *N. sitophila* has been determined to have a powerful cellulolytic capability (10) which is comparable to that of *Chaetomium cellulolyticum* (1, 11), a better known cellulose-degrading fungus (5, 6). Additionally, *N. sitophila* has a processing advantage as being one of the faster growing microfungi. With a maximum specific growth rate of 0.40 h^{-1} it has a doubling time which is shorter than that of some bacteria (12). By comparison, the maximum specific growth rates of other common industrial fungi are half (*e.g.*, for *Aspergillus niger*) or even less than a third (*e.g.*, for *Penicillium chrysogenum*) than that of *N. sitophila* (12).

A microbial biomass protein production process based on *N. sitophila* has been developed and scaled-up to 1300 L (11, 13). The bioprocessing scheme utilizes a number of general unit operations (14). Among these, the two critically important process steps are the size reduction of the cellulosic residue by milling or grinding, and pretreatment of the residue to remove the unwanted lignin and hemicellulose prior to fermentation. These two steps have so far not been investigated in detail. Here we report on the effects of the size of solid substrate particles, and of the conditions of pretreatment, on protein production by *N. sitophila*. Optimization of the milling and the pretreatment operations is essential to the economic viability of the protein production process because these steps are energy intensive and require expensive chemicals.

2. MATERIALS AND METHODS

2.1 Microorganism and Fermentation

Neurospora sitophila (ATCC 36935) was maintained at 4°C on potato dextrose agar (PDA) slants. Seed cultures were prepared in 250 mL conical flasks containing 50 mL seed medium of the following composition (per litre): glucose, 10.0 g; yeast extract (Diffco), 2.0 g; $(NH_4)_2SO_4$, 0.47 g; urea, 0.86 g; KH_2PO_4, 0.714 g; $MgSO_4{\cdot}7H_2O$, 0.2 g; $CaCl_2$, 0.2 g; $FeCl_3$, 3.2 mg; $ZnSO_4{\cdot}7H_2O$, 4.4 mg; H_3BO_3, 0.114 mg; $(NH_4)_6Mo_7O_{24}{\cdot}4H_2O$, 0.48 mg; $CuSO_4{\cdot}5H_2O$, 0.78 mg; $MnCl_2{\cdot}4H_2O$, 0.144 mg. The pH was adjusted to pH 5.5 after sterilization at 121°C for 30-minutes. Flasks cooled to ambient were inoculated from PDA slants and incubated on a rotary shaker (35°C, 220 rpm, 2-days). The biomass produced was aseptically dispersed in a blender (Waring Commercial Blender 7011, Dynamics Corporation of America, New Hartford, CT) at "low intensity" setting for one minute and a 5 mL portion of this material was used to inoculate 50 mL production medium in 250 mL shake flasks. The production flasks were held on a rotary shaker (35°C, 200 rpm) until desired. At desired times, the flasks were removed from the shaker, rapidly cooled and stored at 4°C if necessary. The flasks were analyzed for total dry solids, crude protein and cellulose. All flasks were run in duplicate and the data were averaged. The production conditions (temperature, pH) used had earlier been established to be optimal for *N. sitophila* culture (1, 11).

The production medium contained no yeast extract or glucose. Instead, ground corn stover (10 gL^{-1}) was the carbon source; other medium components were the same as in the seed medium described earlier. Although the media were supplemented with the full complement of the earlier specified nutrient salts, only ammonium sulfate and phosphates were essential requirement with corn stover, a natural substrate which contains other trace nutrients.

2.2 The Carbon Source

Corn stover (corn stalk and leaves) collected from a farm in Waterloo, Ontario, was ground in a Wiley mill (Thomas-Wiley Laboratory Mill, Model 4, Arthur H. Thomas Company, Philadelphia, PA) to a particle size of ≤ 3 mm. This material was sieved through a 2 mm (12 mesh) sieve and further sieved through a 1 mm (20 mesh) sieve to give three particle size fractions: 2-3 mm, 1-2 mm and < 1 mm. These fractions were either pretreated with sodium hydroxide or used without any further treatment. When pretreated, sodium hydroxide was added to corn stover at one of three concentrations of sodium hydroxide (kg NaOH/kg substrate): 0.15, 0.10 or 0.05. The mixture was autoclaved (121° C, 30-minutes) and cooled to ambient. The cooled slurry was washed with 15-20 volumes of deionized water until the pH of the wash became neutral. The intention was to modify the structure of the substrate by removal of some lignin and hemicellulose. The treated corn stover was dried (95° C, overnight) and crumbled prior to formulation in the media.

In one experiment a slightly different pretreatment scheme was employed: ground corn stover (< 1 mm) was autoclaved with 0.10 kg NaOH/kg substrate at the conditions specified above, the pH was adjusted after cooling to room temperature, but the deionized water wash was omitted. The remaining nutrients were added (see above), the material was autoclaved and used as described under Section 2.1.

2.3 Crude Protein and Cellulose

For crude protein and cellulose determinations, the fermentation broth was filtered under suction through a 25 μm "Nitex" nylon cloth (Thomson Co., Scarbrough, Ontario), the filter cake was washed with several broth volumes of deionized water and dried overnight at 90° C. The dry biomass was ground to ≤ 1 mm and a portion was analyzed for total nitrogen using a micro-Kjeldahl technique (15). The crude protein content of the biomass were calculated as 6.25 × total nitrogen, and percent (w/w) protein as gram protein per 100 g total dry solids. The cellulose contents were determined by the spectrophotometric anthrone-sulfuric acid method (16); percent cellulose was calculated on the same basis as crude protein.

3. RESULTS AND DISCUSSION

Utilization of cellulose as a function of fermentation time is shown in Figure 1 for the 0-1 mm corn stover particles treated in various ways. Both the rate of cellulose consumption by the fungus and the overall extent of its utilization were enhanced with increasing concentration of alkali in the pretreatment step (Figure 1). The maximum final cellulose utilization was 28-77% higher (depending on the pretreatment) with the treated substrate relative to the untreated base case (Figure 1). For the best case, which used 0.15 kg NaOH/kg substrate for pretreatment, the final cellulose utilization exceeded *ca.* 90% of the initial cellulose. The improvements in the rate and the extent of cellulose degradation with alkali pretreatment were associated with improvements in accessibility of the cellulases to cellulose which resulted from increasing delignification and hemicellulose removal. The results (Figure 1) implied that even in the 0-1 mm particle size range, the restricted physical access of the enzyme to the substrate was a limiting factor in degradation of cellulose in solid substrates.

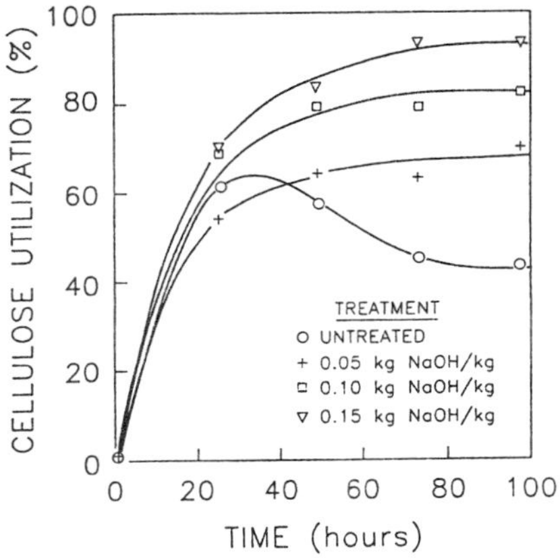

Figure 1. Effect of different pretreatments on cellulose utilization during fermentation of < 1 mm substrate particles.

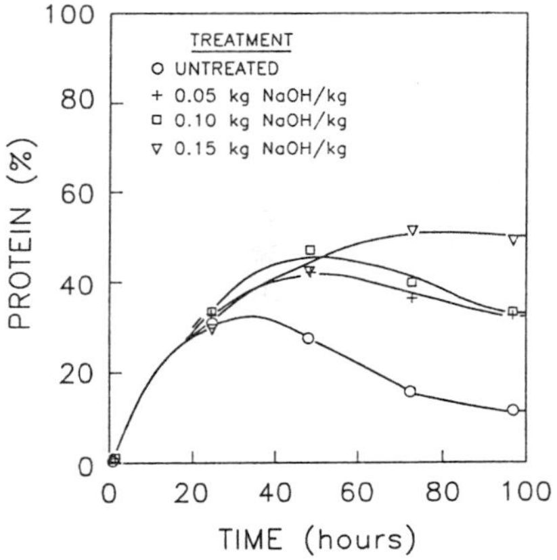

Figure 2. Effect of sodium hydroxide pretreatment on crude protein production during fermentation of < 1 mm substrate particles.

With the untreated substrate, the high apparent cellulose utilization during early part of the fermentation (Figure 1) was an artefact of the measurement method. While the total available cellulose was determined on an alkali treated sample (0.15 kg NaOH per kg substrate) of the substrate and, hence, was the same for all fermentations, low cellulose

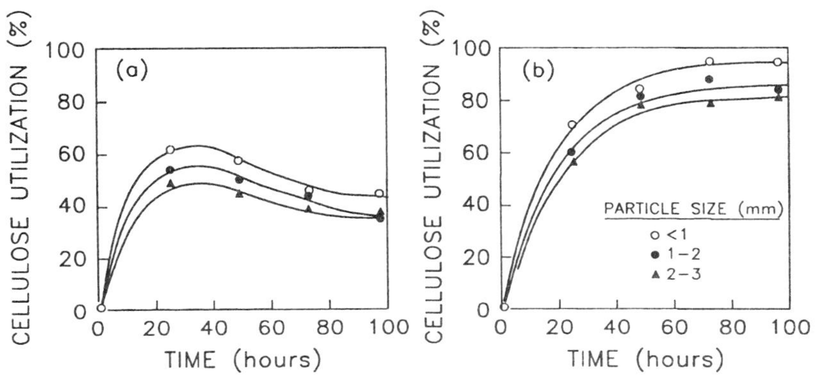

Figure 3. Cellulose utilization as a function of fermentation time for various sizes of solid substrate particles: (a) no pretreatment; (b) pretreated with 0.15 kg NaOH/kg substrate.

values were apparently measured with the untreated substrate possibly because the presence of carbohydrates and lignin hid some of the cellulose from measurement in the untreated material. As the fermentation progressed and the substrate structure opened up, more of the remaining cellulose became measurable. Thus, for untreated particles, the cellulose utilization values obtained toward the end of the fermentation better reflected the actual utilization of cellulose than did data obtained during the 20-60 hour period.

The final yield of the crude protein increased with increasing concentration of alkali during pretreatment as shown in Figure 2 for 0-1 mm substrate particles; however, the initial production rate was not affected very much (Figure 2). This meant that although the cellulose solubilization rates were enhanced (Figure 1), further breakdown of soluble polysaccharides to simple sugars was limiting the growth rate. Nevertheless, for the best case (Figure 2), crude protein made up *ca.* 50% (by weight) of the dry fermentation product within *ca.* 75 hours of fermentation. When soluble carbon sources, such as glucose or molasses, were used the maximum specific growth rate of *N. sitophila* approached *ca.* 0.40 h^{-1} (1, 11, 12) and, starting with the same conditions as used in this work, a 50% protein buildup was achieved in less than 75 hours (1). This was further evidence for limited availability of assimilable sugars as the cause of growth limitation in the solid-slurry fermentations.

The effect of size of the solid substrate particles on cellulose utilization is shown in Figure 3. Although size reduction noticeably improved cellulose

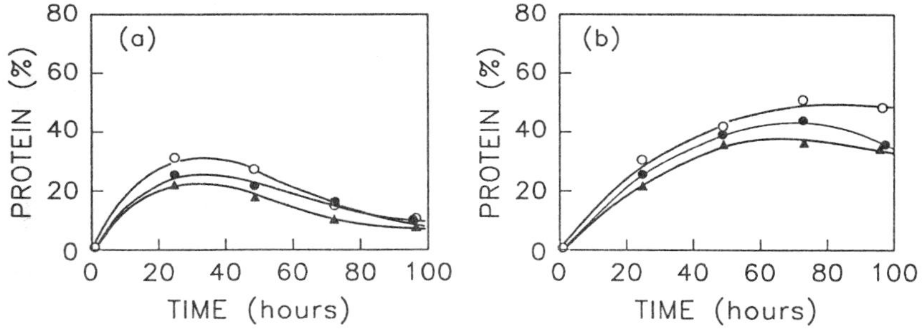

Figure 4. Crude protein production as a function of fermentation time for < 1 mm (O), 1-2 mm (●) and 2-3 mm (▲) solid substrate particles: (a) no pretreatment; (b) pretreated with 0.15 kg NaOH/kg substrate.

degradation in untreated (Figure 3a) as well as alkali-pretreated (Figure 3b) particles, this effect was small in the 0-3 mm size range in comparison with the effect of alkali pretreatment (0.15 kg NaOH/kg substrate). Thus, for the milled and alkali treated particles (Figure 3b), the maximal cellulose utilization was only *ca.* 16% higher for the 0-1 mm particles in comparison with the 2-3 mm particles. On the other hand, even for the larger 2-3 mm particles, alkali pretreatment with 0.15 kg NaOH/kg substrate enhanced cellulose utilization by *ca.* 98% relative to the untreated case (Figure 3). These results implied that in comparison with milling, the alkali-pretreatment was more effective in opening up the solid substrate particles to cellulases. For the untreated substrate, higher than actual cellulose utilization was measured during the early part of fermentation (Figure 3a). This behaviour has been explained earlier for Figure 1.

The crude protein as a function of fermentation time for substrate particles of various sizes, with and without alkali-pretreatment, is shown in Figure 4. The protein production, in particular the final yield, was improved by particle size reduction (Figure 4b) as well as by alkali pretreatment. The decline in the final protein yield with fermentation time (Figure 4a) for the untreated substrate particles arose because by *ca.* 35 hours of fermentation all the readily accessible cellulose had been converted to sugars, lignin and hemicellulose limited further breakdown of the cellulose and the protein yield declined due to starvation associated lysis of the biomass. Once again, although particle size reduction on the untreated substrate enhanced protein yield, the enhancement was not as much as could be achieved by treatment with 0.15 kg NaOH/kg substrate (Figure 4). Earlier observations had indicated that the protein production could potentially be enhanced by improving the accessibility of the substrate to the fungus (11). Substrate availability was believed to be limited either by restricted physical access of the fungal cellulases to the solid particle and/or by inherent limitations in the rate of hydrolysis or solubilization of cellulose (1, 11). In view of the results of this study, the kinetics of the hydrolytic reaction did not seem to be a limiting factor. Note that the secretion of *N. sitophila* cellulases and their inherent hydrolytic capability combined, have already been shown to be comparable to those of other such cellulolytic microfungi as *Chaetomium cellulolyticum* (1, 11) and *Trichoderma reesei* (6).

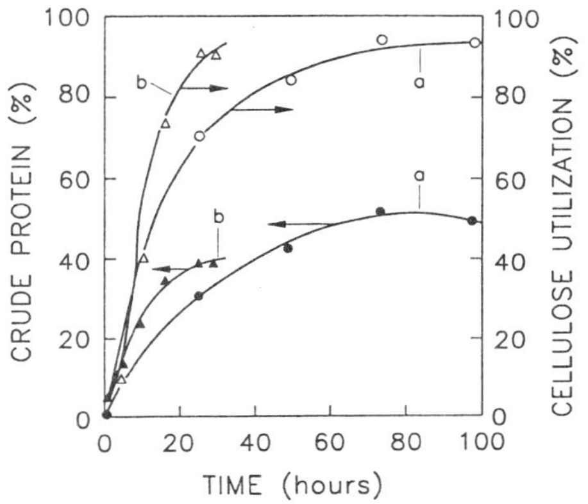

Figure 5. Cellulose utilization and crude protein production during fermentation of < 1 mm substrate particles: (a) pretreated with 0.15 kg NaOH/kg substrate; (b) pretreated with 0.10 kg NaOH/kg substrate using a modified scheme omitting the wash step.

Among the pretreatment options, the omission of the water wash step following alkali treatment reduced handling, simplified pretreatment, reduced water consumption and potential water pollution problems at larger scales; however, the products of degradation of lignin and hemicellulose and any soluble sugars resulting from alkaline-hydrolysis of cellulose, remained in the treated substrate. As shown in Figure 5, the breakdown components of lignin and hemicellulose had no adverse effect on cellulose consumption or protein production by the fungus. In fact, the cellulose utilization and protein production were significantly enhanced in comparison with data obtained on corn stover which had been washed after the alkali pretreatment. This enhanced performance was associated with the initial availability of soluble sugars to the microorganism grown on the unwashed substrate. Note that the pretreatment omitting the water wash utilized a lower concentration of alkali (0.10 kg NaOH/kg substrate) than the 0.15 kg NaOH/kg substrate used with the washed substrate (Figure 5). Clearly, the water wash eliminated the beneficial effects of higher alkali concentration during pretreatment. The fermentation omitting the wash step (Figure 5) was conducted at 37° C versus the 35° C used in all other experiments reported here. In view of the known effects of temperature on *N. sitophila* fermentations (1, 11), the marginally higher temperature did not explain the dramatic improvements in cellulose degradation and protein production which were observed with the alkali treated, unwashed substrate.

4. CONCLUSION

Mycoprotein production with the food-grade fungus *Neurospora sitophila* cultured on ground corn stover was studied. The effect of substrate particle size (0-3 mm) and pretreatment with various concentrations of sodium hydroxide (0-0.15 kg NaOH/kg substrate) were investigated. Both the rate and the overall conversion of cellulose to fungal protein were enhanced by reduction in size of the solid substrate particles as well as by increasing the concentration of sodium hydroxide in the pretreatment step. In the 0-3 mm particle size range, the impact of milling on improving cellulose utilization was relatively less than that of hydroxide pretreatment with 0.15 kg NaOH/kg substrate, *i.e.*, size reduction caused less dramatic enhancement in cellulose utilization than did alkali pretreatment. It seems, therefore, that for small particles the effect of lignin and hemicellulose in reducing the accessibility of cellulases to cellulose is more pronounced than the particle size-associated reduction in enzyme-substrate interactions. Furthermore, the results confirm that even for small (< 1 mm) cellulosic substrate particles in well agitated slurry fermentations the liquefaction of the substrate by cellulases is limited by steric hinderance or transport effects; reaction kinetics do not seem to be the limitation. Up to *ca.* 90% utilization of cellulose could be achieved under the best conditions when the crude protein concentration reached up to *ca.* 50% (by weight) dry solids.

5. REFERENCES

1. Moo-Young, M., Y. Chisti and D. Vlach. 1992. Fermentative conversion of cellulosic substrates to microbial protein by *Neurospora sitophila*. *Biotechnol. Lett.* 14: 863-868.
2. Joglekar, A. V. and N. G. Karanth. 1984. Studies on cellulase production by a mutant *-Penicillium funiculosum* UV 49. *Biotechnol. Bioeng.* 26: 1079-1084.
3. Macris, B. J. 1984. Enhanced cellulase and β-glucosidase production by a mutant of *Alternaria alternata*. *Biotechnol. Bioeng.* 26: 194-196.
4. Herr, D. 1979. Secretion of cellulase and β-glucosidase by *Trichoderma viride* ITCC-1433 in submerged culture on different substrates. *Biotechnol. Bioeng.* 21: 1361-1371.

5. Moo-Young, M., D. S. Chahal and D. Vlach. 1978. Single cell protein from various chemically pretreated wood substrates using *Chaetomium cellulolyticum*. *Biotechnol. Bioeng.* 20: 107-118.
6. Fähnrich, P. and K. Irrgang. 1981. Cellulase and protein production by *Chaetomium cellulolyticum* strains grown on cellulosic substrates. *Biotechnol. Lett.* 3: 201-206.
7. Hesseltine, C. W. and H. L. Wang. 1967. Traditional fermented foods. *Biotechnol. Bioeng.* 9: 275-288.
8. Steinkraus, K. H. 1986. Microbial biomass protein grown on edible substrates: The indigenous fermented foods. In: *Microbial Biomass Proteins* (Moo-Young, M. and K. F. Gregory, ed.), pp. 33-45, Elsevier, London.
9. Wood, J. B. and F. M. Yong. 1975. Oriental food fermentations. In: *The Filamentous Fungi*, vol. **1** (Smith, J. E. and D. R. Berry, ed.), pp. 265-280, Edward Arnold, London.
10. Oguntimein, G., D. Vlach and M. Moo-Young. 1992. Production of cellulolytic enzymes by *Neurospora sitophila* grown on cellulosic materials. *Bioresource Technology* 39: 277-283.
11. Moo-Young, M., Y. Chisti and D. Vlach. 1993. Fermentation of cellulosic materials to mycoprotein foods. *Biotechnol. Adv.* 11: 469-479.
12. Solomons, G. L. 1975. Submerged culture production of mycelial biomass. In: *The Filamentous Fungi*, vol. **1** (Smith, J. E. and D. R. Berry, ed.), pp. 249-264, Edward Arnold, London.
13. Moo-Young, M., R. E. Burrell and J. Michaelides. 1990. Process for upgrading cereal milling by-products into protein-rich food products. *US Patent 4,938,972*.
14. Chisti, Y. and M. Moo-Young. 1991. Fermentation technology, bioprocessing, scale-up and manufacture. In: *Biotechnology: The Science and the Business* (Moses, V. and R. E. Cape, ed.), pp. 167-209, Harwood Academic Publishers, New York.
15. Lang, C. A. 1958. Simple microdetermination of Kjeldahl nitrogen in biological materials. *Anal. Chem.* 30: 1692-1694.
16. Updegraff, D. M. 1969. Semimicro determination of cellulose in biological materials. *Anal. Chem.* 32: 420-424.

ADSORPTION KINETICS OF LYSOZYME ON THE CATION EXCHANGER FRACTOGEL TSK SP-650(M)

M.A. Hashim, K.H. Chu and P.S. Tsan
Institute of Advanced Studies, University of Malaya, Malaysia

Adsorption of lysozyme to the cation exchanger Fractogel TSK SP-650(M) is presented in this paper. The adsorption equilibrium of lysozyme is described by the Langmuir isotherm and the batch dynamic experiments are analyzed using a simplified rate model. The rate parameter determined from different experiments shows no dependence on the experimental conditions.

1. INTRODUCTION

Large-scale protein purification is often carried out by chromatographic techniques which include ion exchange, affinity, size exclusion, reversed phase and hydrophobic interaction chromatography. Of all of these methods of purification, ion exchange adsorbents have found widespread use in both the laboratory and the production plant. A survey carried out by Bonnerjea *et al.* (1) showed that ion exchangers were used in 75% of all the published purification protocols.

To design and optimize purification processes based on ion exchange adsorbents, information concerning the equilibrium and kinetic behaviour of proteins on the ion exchangers is required (Yamamoto et al. (2)). Although extensive research has been conducted on protein adsorption onto various ion exchangers, most of the published papers in this field have dealt mainly with equilibrium behaviour (Huang and Horvath (3); Huang *et al.* (4); James and Do (5)). Thus, little is known regarding kinetic effects on the performance of ion exchange processes.

In this work we evaluated the adsorption kinetics of lysozyme on the cation exchanger Fractogel TSK SP-650(M) through a series of batch adsorption experiments. The kinetic and equilibrium parameters reported here may then be used as the basis for developing methods to analyze adsorption of the protein to a packed bed of Fractogel TSK SP-650(M).

2. THEORY

Consider a vessel which initially contains only a protein solution at a concentration c_o. At time zero, adsorbents which are initially free of protein are added to the vessel. A material balance on the protein in the bulk liquid phase is then given by the following equation.

$$V\frac{dc}{dt} = -v\frac{dq}{dt} \tag{1}$$

where c is the liquid phase protein concentration, q is the adsorbent concentration, V is the liquid volume, v is the adsorbent volume and t is the time. Adsorption of the protein to the ion exchanger is assumed to be monovalent and homogeneous according to the following elementary reaction.

$$P + A \underset{k_2}{\overset{k_1}{\rightleftharpoons}} P \cdot A \tag{2}$$

where P represents the protein molecule, A represents the adsorption site on the ion exchanger, P·A is the protein-ion exchanger complex and k_1 and k_2 refer to the adsorption and desorption rate constants, respectively. The rate of protein adsorption for an interaction described by Eq. (2) is given by

$$\frac{dq}{dt} = k_1 c(q_m - q) - k_2 q \tag{3}$$

where q_m is the maximum binding capacity of the ion exchanger. At equilibrium, Eq. (3) leads to the familiar Langmuir isotherm model.

$$q^* = \frac{q_m c^*}{K_d + c^*} \tag{4}$$

where the superscript asterisk describes an equilibrium value and K_d is a dissociation constant given by the following relation.

$$K_d = \frac{k_2}{k_1} \tag{5}$$

Substituting Eq. (5) into Eq. (3) gives

$$\frac{dq}{dt} = k_1 c(q_m - q) - k_1 K_d q \tag{6}$$

For protein adsorption in a vessel, Eqs. (1) and (6) form the basis for a simplified rate model with a "lumped" adsorption rate constant. The analytical solution of Eqs. (1) and (6) is given by the following equation (Skidmore and Chase (6)).

$$c = c_o - \frac{v}{V}\left[\frac{(b+a)\left[1 - \exp\left(-\frac{2av}{V}k_1 t\right)\right]}{\left(\frac{b+a}{b-a}\right) - \exp\left(-\frac{2av}{V}k_1 t\right)}\right] \tag{7}$$

where

$$a^2 = b^2 - \left[\frac{c_o V}{v}\right] q_m \quad ; \quad b = \frac{1}{2}\left[\frac{c_o V}{v} + q_m + \frac{K_d V}{v}\right] \tag{8}$$

When the values of q_m and K_d are known, the rate parameter k_1 can be determined by fitting the prediction of Eq. (7) to the experimental data of batch adsorption experiments. The remaining rate parameter k_2 can then be calculated from Eq. (5).

3. MATERIALS AND METHODS

Lysozyme was purchased from Sigma (St. Louis, U.S.A.). Acetic acid and sodium acetate were obtained from Fluka (Buchs, Switzerland). 0.1 M acetate buffer solutions were prepared by mixing the required amounts of the chemicals with deionized water. The pH of the buffer solution was maintained at 5.0. The strong cation exchanger, Fractogel TSK SP-650(M), was supplied by E. Merck (Darmstadt, Germany).

3.1 Batch Equilibrium Experiments

A series of protein solutions having different initial concentrations were placed in individual centrifuge tubes with the cation exchanger. The tubes were rotated end-over-end at 25 °C for 24 hours. After equilibration, the slurry was centrifuged, and the protein concentration of the supernatant, c^*, was measured by UV absorbance using a Lambda 3B spectrophotometer (Perkin Elmer, Norwalk, U.S.A.). The amount of protein adsorbed to the ion exchanger, q^*, was calculated by mass balance.

3.2 Batch Kinetics Experiments

The experimental apparatus is shown schematically in Figure 1. The vessel was incubated and agitated in a shaking water bath maintained at 25 °C. A pump (Model 250, Perkin Elmer, Norwalk, U.S.A.) was used to circulate the protein solution in a recycle loop whose volume was kept as small as possible. The ion exchange particles were excluded from the recycle loop by a 10 μm HPLC pump inlet filter. Protein concentration in the liquid phase was monitored continuously using a UV/VIS spectrophotometer equipped with a flow cell (Perkin Elmer, Norwalk, U.S.A.). The absorbance versus time data were recorded by an IBM-compatible PC. Each experiment was carried out in the following manner. The initial absorbance of the protein solution in the vessel was measured, then at time zero, an ion exchange suspension was added to the vessel. The absorbance decreased with time as the protein was being adsorbed, and levelled off as equilibrium was approached. Two adsorption experiments with the same initial protein concentration (1.0 mg/mL) but using different amounts of the ion exchanger (0.5 mL and 1.0 mL) were conducted. The experimental conditions are listed in Table 1.

3.3 Parameter Estimation Procedures

Comparison between the experimental data of the batch equilibrium experiments and model predictions were carried out using a non-linear curve-fitting program in Fig.P (Biosoft, Cambridge, U.K.). A set of best-fit values for the equilibrium parameters q_m and K_d in

Table 1 Experimental Conditions and Values of k_1

Run	c_o (mg mL^{-1})	V (mL)	v (mL)	k_1 (mL mg^{-1} s^{-1})
1	1.0	50	0.5	9.2×10^{-3}
2	1.0	50	1.0	8.8×10^{-3}

Eq. (4) was found. For any given value of the parameter k_1, a unique model prediction of the kinetic studies can now be calculated from Eq. (7) with known values of c_o, V, v, q_m and K_d. The best-fit value of k_1 was estimated by minimizing the error between the theoretical prediction and experimental data. Parameter search was again conducted by the same curve-fitting program.

4. RESULTS AND DISCUSSION

Figure 2 shows the equilibrium isotherm for lysozyme determined for the cation exchanger at 25 °C with a 0.1 M acetate buffer solution of pH 5.0. The equilibrium data are well fitted by the simple Langmuir equation (Eq. (4)). Values of the parameters q_m and K_d estimated by a non-linear curve-fitting program are 38.38 mg/mL and 0.012 mg/mL, respectively. The best isotherm is calculated from these estimates of the parameters and is plotted in solid line in Figure 2.

The dynamic batch adsorption of lysozyme was studied in two experiments using different amounts of the ion exchanger. Figure 3 shows the results of these experiments. It is clear that the adsorption of lysozyme was fairly rapid (equilibrium concentrations were reached in the first 15 minutes). Values of the rate parameter k_1 were estimated by a minimization routine which compared the experimental results with model simulations obtained from Eq. (7). The values of k_1 which gave the best agreement between the experimental data and model predictions are listed in Table 1. These values are in satisfactory agreement, indicating that k_1 shows no dependence on the experimental conditions. The best-fit model simulations are depicted by the solid lines in Figure 3.

Evaluation of adsorption kinetics for a porous adsorbent system is complicated due to the mass transfer processes which occur along with the intrinsic binding kinetics. In this work a simplified rate model which lumped mass transfer resistances and intrinsic binding kinetics into a single rate equation was employed to analyze the adsorption of lysozyme. The rate parameter k_1 may be regarded as the intrinsic kinetic constant in cases where the binding kinetics are rate-limiting. However, the simplified rate model is likely to underestimate the intrinsic kinetic constant when the rate of mass transfer becomes comparable to the rate of binding. The intrinsic kinetic constant may be evaluated from a rigorous rate model which accounts for film mass transport, intraparticle diffusion and binding kinetics (Arve and Liapis (7)). However, this type of analysis requires sophisticated numerical techniques to solve the model equations. In addition, accurate determination of each rate parameter in the rigorous rate model requires elaborate experimental procedures.

5. CONCLUSION

A methodology for the quantitative analysis of a batch protein adsorption process has been presented in this paper. The batch method was employed for the equilibrium and dynamic studies due to its relative simplicity in both the experimental procedures and computation involved in the solution of the simplified rate model. The equilibrium and kinetic parameters determined here may be used in the analysis of adsorption of lysozyme to a packed bed of Fractogel TSK SP-650(M).

SYMBOLS

A	adsorption site of ion exchanger
c	protein concentration in the bulk phase
c_o	initial protein concentration of bulk phase
c^*	protein concentration of bulk phase at equilibrium
k_1	lumped forward rate constant
k_2	lumped reverse rate constant
K_d	apparent dissociation constant
P	molecule of protein
P·A	protein-ion exchanger complex
q	average concentration of adsorbed protein
q_m	maximum binding capacity of ion exchanger
q^*	concentration of adsorbed protein at equilibrium
v	volume of ion exchanger
V	volume of protein solution
t	time

REFERENCES

1. Bonnerjia, J., Oh, S., Hoare, M. and Dunnill, P., 1986, Biotechnology, 4, 954.
2. Yamamoto, S., Nakanishi, K. and Matsuno, R., 1988, Ion-Exchange Chromatography of Proteins, Marcel Dekker, New York.
3. Huang, J.-X. and Horvath, Cs., 1987, J. Chromatogr., 406, 285.
4. Huang, J.-X., Schudel, J. and Guiochon, G., 1991, J. Chromatogr. Sci., 29, 122.
5. James, E.A. and Do, D.D., 1991, J. Chromatogr., 542, 19.
6. Skidmore, G.L. and Chase, H.A., 1988, In Ion Exchange for Industry, M. Streat (Ed.), Ellis Horwood, Chichester, 520.
7. Arve, B.H. and Liapis, A.I., 1987, AIChE J., 33, 179.

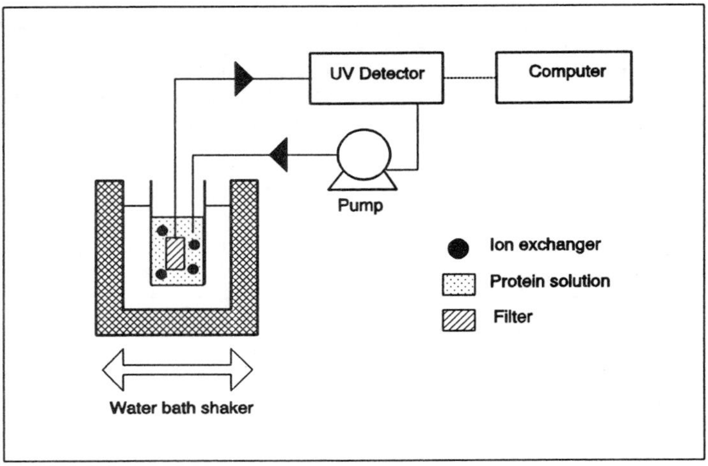

Figure 1 Apparatus for measuring the kinetics of adsorption.

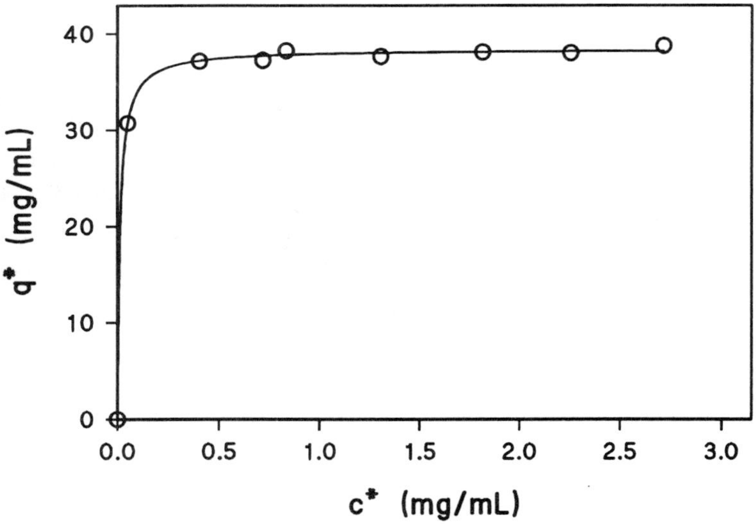

Figure 2 Equilibrium isotherm for binding of lysozyme to Fractogel TSK SP-650(M). The soild line represents the best-fit Langmuir equation.

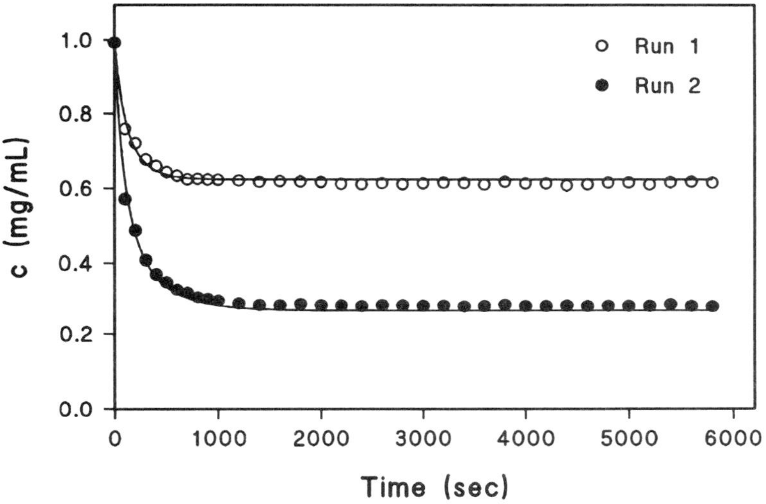

Figure 3 Kinetics of adsorption of lysozyme to Fractogel TSK SP-650(M). The best-fit model simulations are represented by the solid lines.

THE EFFECT OF IMPELLER CONFIGURATION ON BIOLOGICAL PERFORMANCE IN NON-NEWTONIAN FERMENTATIONS

M.J. Kennedy, and R.J. Davies
New Zealand Institute of Industrial Research and Development, Box 31-310, Lower Hutt, New Zealand

ABSTRACT

Non-Newtonian fermentations involving mycelial microorganisms, for example penicillin production, are of considerable commercial significance. While many stuies have looked at the effect of impeller type on engineering parameters, such as oxygen transfer rate, very few studies have investigated the effect of impeller type on the biological productivity of the system, for example product concentration. This study provides pilot plant data on the effect of impeller type on fermentation performance. Under conditions of constant power supplied to a fermenter, the 45° pitched bladed impeller proved to be 7-40% better in terms of biological performance in non-Newtonian fermentations, than the standard Rushton turbine impeller. This was demonstrated for two mycelial fermentations, gamma linolenic acid production by *Mucor hiemalis* IRL 51, and biocontrol agent production by *Truncatella angustata* IRL 167A. Thus non-Newtonian fermentation performance can be significantly improved by using the correct impeller configuration.

1. INTRODUCTION

Mycelial fermentations are of considerable industrial importance. Examples of important mycelial fermentations include the production of antibiotics by *Penicillium* and *Streptomyces* species, the production of α-amylase by *Aspergillus oryzae*, and the production of rennet by *Mucor pusillus*. Many mycelial fermentations are non-Newtonian. A non-Newtonian fluid is one that does not follow Newtonian fluid behavior, ie the viscosity varies with the agitation rate (Walker and Cox, 8). Typically, non-Newtonian fermentations are very viscous. The non-Newtonian behavior of the fluid creates problems with mixing and oxygen transfer during the fermentation.

Traditionally non-Newtonian fermentation systems were agitated with Rushton impellers, named after the pioneering work on mixing conducted by Rushton et al (7), in the 1950s). Later it became known that Rushton impellers were not ideally suited to mixing non-Newtonian fermentations (Nienow, 6). Other impeller designs were experimented with and now many such impellers, designed for non-Newtonian fermentations exist (Buckland et al, 1-2).

These new impeller designs have been extensively studied in terms of mixing capability (Cooke et al, 3) and oxygen transfer capability (Buckland et al, 1-2), but little work has been conducted on the effect of impeller configuration on the biological performance of the fermentation. Biological performance can be measured by such variables as the cell dry weight, the product concentration, or the product volumetric productivity. This study provides experimental data on the effect of impeller configuration on the biological performance of two commercially important non-Newtonian systems; the production of gamma linolenic acid (GLA) by *Mucor hiemalis* IRL 51 and the production of a potential biocontrol agent, *Truncatella angustata* IRL 167a. These studies were conducted on the pilot plant (200L) scale.

2. MATERIALS AND METHODS

2.1 Microorganisms Used

Two filamentous fungi were used for this investigation; *Mucor hiemalis* IPD 51 which produces gamma linolenic acid (GLA), which is a high value pharmaceutical with medical applications; and *Tuncatella angustata* IPD 167a, which is a biocontrol agent being developed to combat silver leaf disease in deciduous trees. IRL refers to the New Zealand Institute of Industrial Research and Development Culture Collection.

2.2 Media

The following medium was used to grow *M. hiemalis* IPD 51:

dextrose monohydrate	160 g/L
casein hydrolysate	20 g/L
mauri yeast extract	3.6 g/L
ppg[+]2000	40 mL/170L
([+]ppg = polypropyleneglycol)	

The following medium was used to grow *T. angustata* IPD 167a:

glucose monohydrate	40 g/L
BM yeast extract	1 g/L
V8 juice	200 mL/L
AF1520 10%	40 mL/170L

2.3 Fermentation Conditions

	M. hiemalis	*T. angustata*
Temperature	25°C	25°C
pH	5.0	5.5
pH controls used	6M NaOH or 4M H_2SO4	6M NaOH/4 or 6M H_2SO4
Working volume	170 L	170 L
Air flowrate	60 L/min	60 L/min
Power input	300 Watts	200 Watts

2.4 Impellers

The following four impellers were investigated in both fermentation systems:

Rushton impeller
Hydrofoil impeller
45° pitch bladed impeller
Curve bladed impeller

The impeller was positioned on the fermenter agitation shaft so that the mid-point of the impeller was 1/2 the way up from the bottom of the fermenter to the top of fermentation liquid. All impellers were 270 mm in diameter.

2.5 Fermenter

The fermenter used was a 200L nominal volume baffled, stirred tank with internal coils for temperature control. The fermenter was run with a liquid height to tank diameter, or aspect ratio, of 1:1. All fermentations were conducted at constant power input to the fermentation broth.

2.6 Torque and Speed Measurement

The Energy Applications team of Industrial Research Limited have developed a novel dynamometer drive coupling that was fitted to the fermenter to measure applied torque and speed of the rotating fermenter shaft (Kennedy et al, 4).

2.7 Power Calculation

Once the torque acting on the impeller, and the impeller speed have been measured then the power input into the fermenter can be calculated using the following formula:

Power (Watts) = Torque (Nm) x Impeller Speed (rpm) x 2 x π / 60.

2.8 Mixing Time Calculation:

Mixing time was measured by monitoring the change in pH after a pulse of concentrated acid or base was added to the fermenter. These measurements were always conducted at the end of each fermentation.

2.9 Cell Concentration, Oil Content, Fatty Acid Profile

Cell concentration, the oil content of the cell and fatty acid composition of the oil were estimated according to standard protocols, the details of which can be found in Kennedy et al (5).

2.10 Viscosity

Viscosity was measured at varying shear rates using a Brookfield Synchro-Lectric viscometer (Brookfield Engineering Laboratories, Stoughton, Massachusetts, USA), model RV, spindle numbers 1 and 2.

3. RESULTS AND DISCUSSION

3.1 Gamma Linolenic Acid Production

For the *Mucor hiemalis* IPD 51 fermentation cell dry weight was taken as the indicative biological performance parameter. This is because in a wide range of fermentation processes, cell dry weight is an important variable. The 45° pitch bladed impeller gave the best result with a 7 to 15% increase in maximum cell dry weight compared to the standard Rushton impeller configuration in the best case (see table 1).

When the average torque produced in the fermentations was plotted against the maximum dry cell weight obtained, the maximum cell dry weight was inversely related to the average torque. This has implications for design purposes. For this fermentation, the impeller that produced the least torque, (or shear) for a constant power input produced the greatest biological performance. The mixing time traces measured at the end of each fermentation were similar, showing that under the conditions studied good mixing times could be maintained regardless of impeller configuration.

Table 1: The effect of impeller geometry on the maximum cell dry weight obtained in the fermentation of *Mucor hiemalis* IPD 51.

Impeller Used	Maximum Cell Dry Weight (g/L)	Relativity to Rushton (%)
Rushton	39.1	100
Hydrofoil	38.7	99
45° pitch bladed	44.8	115
Curve bladed	39.3	101
45° pitch bladed repeat run	41.6	107

3.2 Biocontrol Agent Production

For *Truncatella angustata* IPD 167a, the viscosity of the medium was taken as the indicative biological performance parameter. Cell concentration was not chosen because this microorganism produces an exopolysaccharide which makes dry weight techniques unreliable. Viscosity is assumed to be proportional to exopolysaccharide concentration in the fermentation broth, and hence this system is a good example of optimization of an excellular microbial product. A 41% increase in viscosity can be obtained by using the 45° pitch bladed impeller compared to the Rushton impeller (see table 2).

When the average torque produced in the fermentations was plotted against the viscosity an inverse relationship was observed. This same inverse relationship can be seen when torque is plotted against time for glucose exhaustion, or maximum growth rate (from CO_2 data). These results again show that the impeller that produced the least torque (or shear stress) for a constant power input gave the best biological performance. Again as with *Mucor hiemalis* IPD 51, there was no major differences in mixing time seen between the different impellers studied.

Table 2: The effect of impeller geometry on the viscosity obtained in the fermentation of *Truncatella angustata* IPD 167a.

Impeller Used	Viscosity at 0.5 rpm (cP)	Relativity to Rushton (%)
Rushton	11,800	100
Hydrofoil	16,400	139
45° pitch bladed	16,600	141
Curve bladed	14,600	124

3.3 Comparisons

It is interesting to note that in both fermentations the 45° pitch bladed impeller gave the best results, and in both cases the least torque produced the better result. In all fermentations the mixing times were approximately equal with no impeller effect at the power levels used.

4. CONCLUSIONS

4.1 For both cell mass accumulation by *Mucor hiemalis* IPD 51 and viscosity development by *Truncatella angustata* IPD 167a, the 45° pitch bladed impeller gave the best result with a 7-41% improvement over the Rushton impeller.

4.2 There is an inverse relationship between the torque produced in the fermentation and the relevant biological indicators, at constant power input to the fermentation.

4.3 The data in this study imply that the lower the torque level that an impeller design can produce, the better the biological performance will be, assuming that the mixing time is not significantly affected.

5. REFERENCES

1 Buckland, B.C., Gbewonyo, K., DiMasi, D., Hunt, G., Wasterfield, G., and Nienow, A.W., (1988) Biotechnology and Bioengineering, 31(7), 737-742.

2 Buckland, B.C., Gbewonyo, K., Jain, D., Glazomitsky, K., Hunt, G., and Drew, S.W., (1988) Bioreactor Fluid Dynamics, 1-15.

3 Cooke, M., Middleton, J.C., and Bush, J.R., (1988) Bioreactor Fluid Dynamics, 37-64.

4 Kennedy, M.J., Davies, R.J., Patrick, A., Reader, S.L., and Swierczynski, L., (1994) Industrial Research Limited Reports Nos. 53010, 53012.

5 Kennedy, M.J., Reader, S.L., and Davies, R.J., (1993) Biotechnology and Bioengineering, 42, 625-634.

6 Nienow, A.W., (1990) Trends in Biotechnology, 8(8), 224-233.

7 Rushton, J.H., Costick, E.W., and Everett, H.J., (1950) Chem. Eng. Progress, 46, 467-476.

8 Walker, J.M., and Cox, M., (1988) The Language of Biotechnology, A Dictionary of Terms, American Chemical Society, Washington DC.

PREPARATION AND CHARACTERIZATION OF ACTIVATED CARBON DERIVED FROM PALM OIL SHELLS USING A FIXED BED PYROLYSER

Normah Mulop[1], K.C. Teo[2] and A.P. Watkinson[2]
[1]Department of Chemical Engineering, Universiti Teknologi Malaysia, Kuala Lumpur, MALAYSIA and
[2]Department of Chemical Engineering, University of British Columbia, Vancouver, B.C., CANADA

The preparation of activated carbon from palm oil shells was carried out in two consecutive steps: carbonization of the raw material at 450 °C to the intermediate char which was converted to the activated carbon product via steam gasification at 850 °C. For every different variable under study the yield and the specific surface area of the product were determined. The optimum duration of activation was 45 minutes which gave a specific surface area of 710m^2/g and a yield of 21 %.

1.0 INTRODUCTION

The purpose of this study is to prepare and characterize granular activated carbon derived from palm oil shell which is abundantly available as a solid waste from palm oil production operations. Although most of the shell is being fed to the mills' boilers, there is still a large surplus. This material can be converted to various grades of activated carbon, which are of great demand from the regional processing industry for purification of wastewater and drinking water.

The first step in the activated carbon preparation is a carbonization process in an inert atmosphere. This is followed by a controlled steam gasification for the activation process. The carbonization temperature, the duration of carbonization as well as the temperature of activation are held constant throughout this study. The duration of activation, and the amount and injection rate of water are studied.

2.0 EXPERIMENTAL

All experiments were carried out in an atmospheric pressure fixed-bed batch operation in a 500 mL stainless steel pyrolyser. The reactor was purged with a constant flow rate of nitrogen continuously during both the carbonization and activation steps. The nitrogen flow rate was controlled in order to avoid the entrainment of the activated carbon. Carbonization temperature and steam activation temperature were held constant at 450 °C and 850 °C respectiely. The duration of carbonization carried out was

60 minutes. Various combinations of water injection (feed) rate and duration of steam activation were investigated.

The vapour derived from the carbonization step was trapped and collected as tar-water liquid in a series of condensers. Recovery was closed to quantitative yield and was compared to the sum of moisture and volatile contents obtained from thermogravimetric analysis (TGA). A Perkin-Elmer TGA-2 Analyser equipped with data station was used for the quantitative analysis of the palm oil shell. Surface area was determined by single-point B.E.T method using nitrogen absorbate in helium carrier gas on a Micromeritic Surface-Analyser (model 2300).

3.0 RESULT AND DISCUSSION

3.1 Water Injection Rate

At high water injection rate, the activated carbon produced would be expected of a lower yield with a higher specific surface area.

The yield of activated carbon decreased with increase in the ratio of gram of water feed per gram of palm oil shell (w). Initially the specific surface area (S.S.A.) increased with 'w' to about 1.0 and then the S.S.A. became almost constant at high 'w' values as shown in Figure 1. Thus, for a fixed duration of activation, there existed an optimum 'w' value in order to achieve the highest S.S.A. For a 30-minute activation, at w = 1.1, one can easily achieve an activated carbon with S.S.A. of 600 m^2/g and a yield of 23 %.

3.2 Duration Of Steam Activation

For a longer duration of activation, a lower yield with a higher S.S.A. product would be expected.

In Figure 2 the relationship between the yield and the duration of activation agreed with the expected result. The yield of activated carbon decreased with the duration of activation and a reduction in yield gave a higher S.S.A. Although a longer duration of activation gave a higher S.S.A., after a duration of 45 minutes the S.S.A. showed a slight decrease or became constant. Thus the optimum duration of activation obtained was 45 minutes which gave a S.S.A. of 712 m^2/g and a yield of 21 %.

3.3 Comparison With Previous Result

Comparison of yield of activated carbon and specific surface area versus duration of activation for various raw materials such as palm oil shell, sawdust, scrap tires and lignin was shown in Figures 3 and 4. However, other than for palm oil shell, the duration of activation for the other raw materials were short and varied between 5 to 18 minutes. The carbonization temperature, duration of carbonization and the activation temperature were also different for most of the other raw materials.

It is obvious that all these materials followed a similar trend, that the yield decreased with the duration of activation. The yield of activated carbon from scrap tires were the highest. This might be due to the short carbonization time of 3 minutes. Sawdust carbonised by Normah (1) gave the lowest yield compared to the sawdust carbonised by Jensen (2). The lower yield might be due to the combination of a higher activation temperature and steam injection rate. Lignin gave a very steep reduction in yield when the duration of activation was varied from 6 to 18 minutes. The chemical properties of the material might have some influence on the yield. Materials with a high fixed carbon content would be expected to give a higher activated carbon yield.

Activated carbon from palm oil shell gave the highest specific surface area followed by that from sawdust (Normah (1)), lignin (Lai (3)), sawdust (Jensen (2)) and scrap tires Kan (4). However, the result was difficult to compare with this study, due to shorter duration of activation for most of the raw materials as well as significant difference in some of the carbonization conditions.

3.4 Tar Yield

The percent tar yield (including moisture) obtained during the carbonization process of palm oil shell was 56 % ($\pm$5%). The average tar yield for palm oil shell agreed well with the result obtained from thermogravimetric analysis (TGA).

3.5 Thermogravimetric Analysis

Figure 5 shows the TGA proximate analysis of raw palm oil shell. From this TGA study a 32 % char yield (i.e. fixed carbon content) was found at an isothermal temperature of 450 ^{0}C, which was in good agreement with the 34 % char yield obtained from the bench-scale carbonization experiments at the same temperature (Table 1). This demonstrated that the carbonization conditions in the pyrolyser were well optimised for maximum yield of char and tar liquid. Finally, the result obtained from pyrolyser equipment was reproducibe.

4.0 CONCLUSION

(1) Increasing the duration of activation decreased the yield of activated carbon from palm oil shell with an increase in specific surface area.

(2) The specific surface area increased with the ratio of gram of water feed per gram of palm oil shell used, up to a value at 1.1, then became almost constant at a higher value of the ratio.

(3) All the biomass based raw materials followed a similar trend in that the yield of activated carbon decreased while the specific surface area increased with the duration of activation.

(4) Activated carbon from palm oil shell gave the highest specific surface area among all biomass based material tested under similar processing conditions.

TABLE 1: Effect of Processing Conditions on the Yield and Specific Surface Area of Activated Carbon

Experiment No.	1	2	3	4	5	6	7
Duration of activation, min*	0	15	25	30	45	60	60
Amount of water feed/g shell	0	0.2	0.2	0.7	1.5	1.5	1.5
Percent yield, %	34.2	26.7	23.6	23.8	20.9	18.8	19.3
Specific surface area, m^2/g	5.2	203	461	557	713	679	709

Other process parameters: Temperature of carbonization: 450 ^{o}C
[Duration = 60 minutes]
* Temperature of activation: 850 ^{o}C

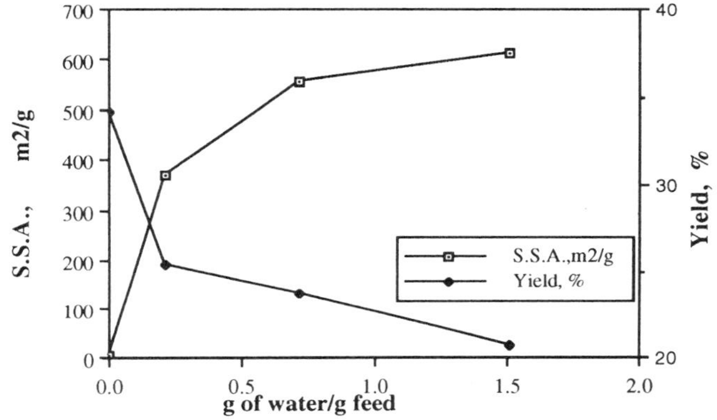

FIGURE 1: SPECIFIC SURFACE AREA AND YIELD VERSUS TOTAL GRAM OF WATER/GRAM FEED

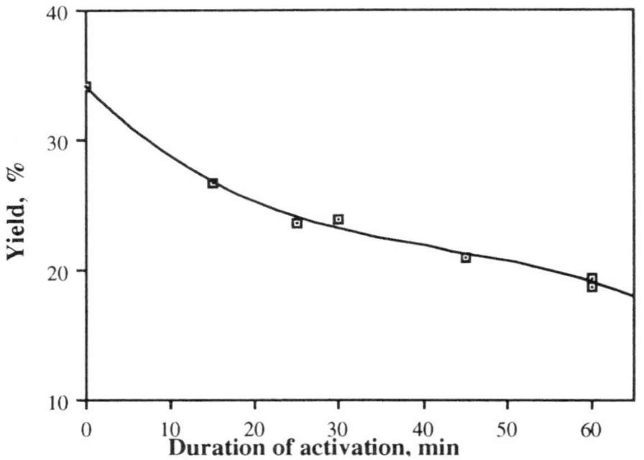

FIGURE 2: YIELD OF ACTIVATED CARBON VERSUS DURATION OF ACTIVATION

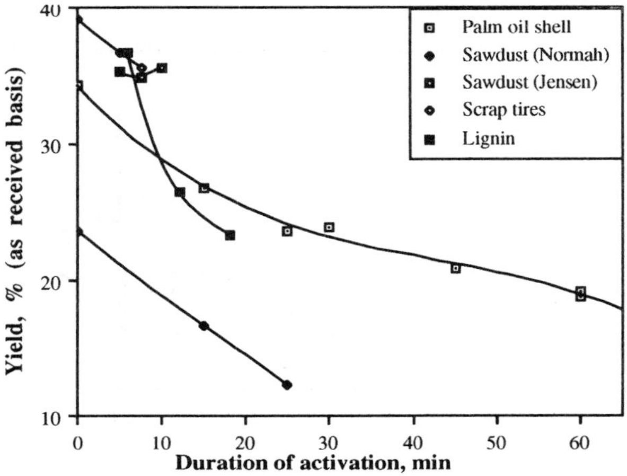

FIGURE 3: COMPARISON OF YIELD VERSUS DURATION OF ACTIVATION FOR VARIOUS RAW MATERIAL

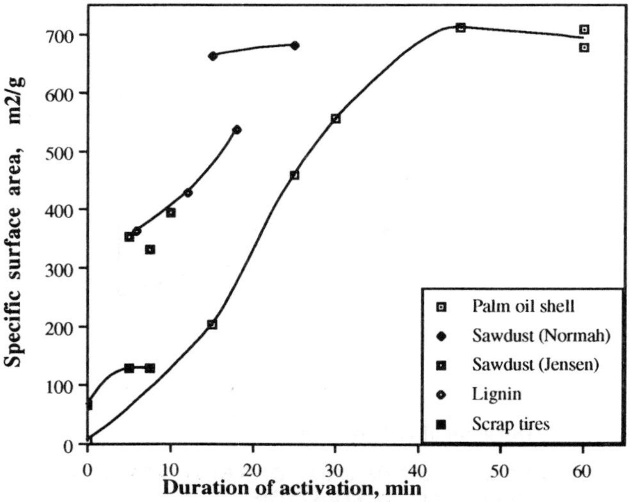

FIGURE 4: COMPARISON OF SPECIFIC SURFACE AREA VERSUS DURATION OF ACTIVATION FOR VARIOUS RAW MATERIAL

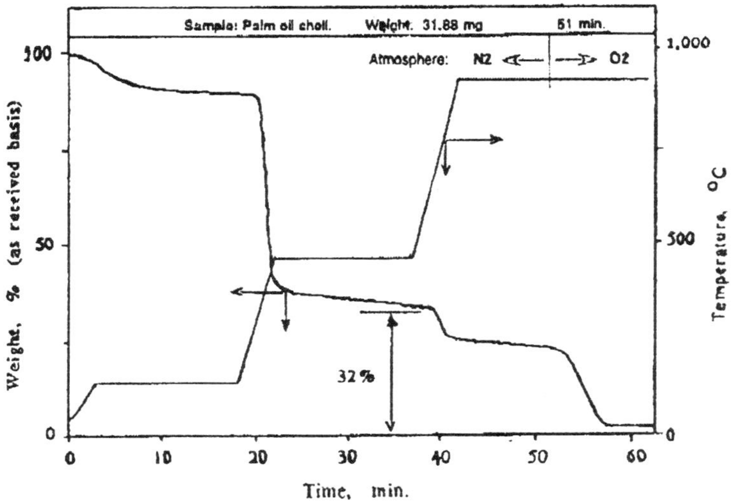

FIGURE 5: TGA OUTPUT OF PALM OIL SHELL

5.0 ACKNOWLEDGEMENTS

Normah Mulop thanks the Universiti Teknologi Malaysia for the sponsorship, the Asean Development Bank for their financial support, and to the Department of Chemical Engineering, University of British Columbia for allowing the authors to carry out the study.

6.0 REFERENCES

1. Normah, M, 1993, "Preparation of activated carbon from palm oil shell using a bench scale pyrolyser", Research report, University of British Columbia.

2. Jensen, A, 1989, "Preparation of activated carbon from British Columbia forest industry waste wood products", B.A.Sc. thesis, University of British Columbia.

3. Lai, B, 1990, "Preparation of activated carbon from Lignin", B.A.Sc. thesis, University of Britih Columbia.

4. Kan, F, 1991, "Preparation of activated carbon from scrap tires", B.A.Sc. thesis, University of Britih Columbia.

5. Winguist, R E., 1992 "Porosity of activated carbon", B.A.Sc. thesis, University of British Columbia.

ROLES OF BACTERIA IN ANAEROBIOSIS OF BRANCHED-CHAIN FATTY ACIDS

H. Chua[1*], M.G.S. Yap[2] and W.J. Ng[2]

[1]CSE, Hong Kong Polytechnic University, Kowloon, HONG KONG
[2]National University of Singapore, Kent Ridge Crescent, SINGAPORE

This paper identifies the bacteria and their roles in the anaerobiosis of branched-chain fatty acids (BCFA), which are persistent and present in pharmaceutical wastes. The consortium of acid-degrading *Syntrophomonas spp.*, H_2-utilizing *Methanococcus spp.* and acetoclastic *Methanothrix spp.* could degrade BCFAs with tertiary carbon but not those with quaternary carbon at the alpha or beta position. The fate of BCFAs in the degradation by anaerobic consortium was elucidated and beta-oxidation was proposed as the mechanism of acidogenesis.

1. INTRODUCTION

Biodegradation of BCFAs has not been intensively studied because these compounds are generally absent in common domestic wastes. In recent years, BCFAs such as 2-methylbutanoic acid (2-MBA) and 3-methybutanoic acid (3-MBA) have been found to be produced through anaerobic degradation of a number of proteinaceous wastes (Masey *et al* (1)); 2,2-dimethylpropanoic acid (2,2-DMPA) and 2-ethylhexanoic acid (2-EHA) are discharged by the pharmaceutical industry (Ng *et al* (2)). These BCFAs are persistent in on-site aerobic treatment facilities and cause adverse impact on the environment. Recent studies by Chua *et al* (3) and Jimeno *et al* (4) showed that these BCFAs are biodegradable in anaerobic filters. Richardson *et al* (5) isolated a 2-MBA-dagrading anaerobic consortium, comprising of an obligate syntroph and two methanogens, but the degradation mechanism was not described.

This paper identifies the bacteria and their roles in the anaerobiosis of 2-EHA and other BCFAs, and elucidates the degradation pathway and mechanisms, on which a mathematical model can be based.

2. MATERIALS AND METHODS

2.1 Enrichment Cultures

The mixed liquor from an anaerobic reactor, which had been treating 2-EHA for three years, was used as inoculated medium in enrichment cultures. The inoculated medium was kept under anaerobic condition for a week to utilize the residual organic components, and then supplemented with minerals and a growth factor.

Two series of cultures were prepared in 100-mL serum bottles with different fatty acids as the sole carbon source (Table 1). Each culture from the first series was periodically fed with increasing acid concentration until the point of inhibition was reached. The cultures were prepared and maintained at 37°C in an anaerobic chamber (Forma Scientific Inc. Model 1029).

2.2 Bacterial Observation and Enumeration

A scanning electron microscope (Joel JSM-T220A) was used to observe and enumerate the bacteria. Samples were prepared with techniques similar to those described by Drier *et al* (6). One-mL sample from the culture was filtered through a Nuclepore (13 mm x 0.4 micron) cellulose nitrate membrane, which was pre-wet with Triton-X 100 fluid to ensure uniform distribution of bacteria on the membrane. The fixed sample was coated with 25 nm gold-palladium (Joel Fine-Coat Ion Sputter Type JFC-1100) and observed at 10 kV accelerating voltage and 5,000-20,000 times magnification.

The bacterial cell density was calculated as $X(A_1/A_2)/v$, where X was the number of cells seen on the micrograph, A_1 and A_2 were the areas of filter membrane and field of micrograph respectively, and v was the volume of sample.

Table 1 - Fatty Acids in the Enrichment Cultures

SERIES 1 Carbon source	SERIES 1 Initial conc.	SERIES 2 Carbon source	SERIES 2 Initial conc.
Ethanoic acid	250	2-Methylbutanoic acid (2-MBA)	200
Butanoic acid	150	2-Methylpropanoic acid (2-MPA)	200
2-Ethylhexanoic (2-EHA)	200	3-Methylbutanoic acid (3-MBA)	200
Propanoic acid	50	2-Ethylbutanoic acid (2-EBA)	200
		2,2-Dimethylpropanoic acid (2,2-DMPA)	200
		2,2-Dimethylbutanoic acid (2,2-DMBA)	200
		3,3-Dimethylbutanoic acid (3,3-DMBA)	200

2.3 Analytical Methods

Fatty acid concentrations were analysed by a Chromosorb WAW 100/120 mesh (FFAP 15% and H_3PO_4 1%) gas-chromatographic column. Biogas quality was examined using a 2 m Porapak Q 80/100 mesh gas-chromatographic column.

3. RESULTS AND DISCUSSION

3.1 Ethanoic-Acid Enrichment

A monoculture of filamentous rods with distinctive truncated ends was isolated in the ethanoic enrichment (Plate 1). These cells (0.3-0.8 x 3-15 microns) had no discernible cross wall and were morphologically similar to the ethanoate-utilizing *Methanothrix spp.*. Presence of *Methanothrix* as the sole ethanoate utilizer was consistent with the results reported by Huser *et al* (7) and Zehnder *et al* (8) that the *Methanothrix* has a higher affinity towards ethanoic acid than other acetoclastic methanogens, namely the *Methanosarcina spp.*.

The monoculture, which reached a maximum density of 1.1 x 10^8 cells/mL, degraded ethanoic acid at a maximum specific degradation rate of 4.9 x 10^{-10} mg/h -cell, and was completely inhibited when ethanoic acid concentration exceeded 2,900 mg/L.

3.2 Butanoic-Acid and 2-Ethylhexanoic-Acid Enrichments

The bacteria consortia in the butanoic-acid and 2-EHA enrichments (Plates 2 and 3 respectively) were similar and comprised of three morphologically distinctive species. These were (1) rods (0.3-2.0 x 1.5-5.0 microns), (2) cocci (0.5-1.2 microns diameter) which fluoresced when excited at 420 nm, and (3) rods which were similar to those observed in the ethanoic-acid enrichment. The first group of rods were morphologically similar to the acid-degrading *Syntrophomonas spp.* described by McInerney *et al* (9,10). The fluorescent property of the cocci was attributed to the intracellular coenzyme F_{420nm}, thus identifying these cells as the H_2-utilizing *Methanococcus spp.* (Jones *et al* (11)). The presence of *Methanococci* demonstrated the obligate syntrophic behavior of *Syntrophomonas* in interspecies hydrogen transfer.

The acid concentration profiles in the butanoic-acid and 2-EHA enrichments are shown in Figures 1 and 2 respectively. Detection of ethanoic acid in the butanoic-acid enrichment, and butanoic and ethanoic acids in the 2-EHA enrichment suggested that 2-EHA was beta-oxidized to ethanoic acid by cleavages between the alpha- and beta-carbon along the main chain of the 2-EHA molecule (Figure 3). Cleavage at the ethyl side chain would have otherwise produced hexanoic acid as an intermediate product. Figure 4 shows the fate of 2-EHA in anaerobiosis. 2-EHA was beta-oxidized by the *Syntrophomonas*, via butanoic acid, to ethanoic acid with concomitant H_2 production. Ethanoic acid was decarboxylated by the *Methanothrix* to CH_4 and CO_2 while H_2 was utilized by the *Methanococci* to reduce CO_2 to CH_4.

The butanoic-acid culture, which reached a maximum density of 1.9 x 10^7 cells/mL, degraded butanoic acid at a maximum specific degradation rate of 4.8 x 10^{-10} mg/h -cell and the culture was completely inhibited when butanoic acid concentration exceeded 1,000 mg/L. The 2-EHA culture, which reached a maximum density of 4.2×10^7 cells/mL, degraded 2-EHA at a maximum specific degradation rate of 2.7×10^{-10} mg/h -cell and the culture was completely inhibited when 2-EHA concentration exceeded 2,200 mg/L. These results showed that the degradation rate was higher with short-chain fatty acids. However, butanoic acid appeared to be most inhibitory to the anaerobic consortium.

3.3 Propanoic-Acid Enrichment

Propanoic acid as the sole carbon did not sustain the consortium, indicating that *Syntrophomonas spp*. could not effectively utilize propanoate as a carbon source.

3.4 Other Branched-Chain Fatty-Acid Enrichments

Branched-chain fatty acids such as 2-MBA, 2-MPA, 3-MBA and 2-EBA have the methyl or ethyl substituent at the alpha- or beta-carbon, resulting in a tertiary carbon, near the carboxylic end of the carbon chain. This characteristic differentiates these compounds from the natural lipid-origin anteiso fatty acids described by Smith (12), which are substituted at the antepenultimate position (third carbon from the alkyl end). The substituents at the alpha- or beta-positions appeared to interfere with the cleaving mechanism in beta-oxidation. The anaerobic consortium in the enrichment cultures was observed to degrade 2-MBA, 2-MPA, 3-MBA and 2-EBA, at initial specific degradation rates around 1.0×10^{-10} mg/h -cell. These results agreed with the findings of Jimeno *et al* (4) and Richardson *et al* (5) that, although being persistent, BCFAs with a tertiary carbon could be degraded by anaerobic consortia isolated through appropriate enrichment techniques and culture conditions. The acid concentration profiles in 2-MBA degradation, which is typical amongst degradable BCFAs, are shown in Figure 5. Propanoic acid was produced through the degradation of 2-MBA, which agreed with the degradation mechanism proposed in Figure 3.

On the other hand, 2,2-DMPA, 2,2-DMBA and 3,3-DMBA were not degradable by the consortium. These recalcitrant BCFAs are different from the those isolated from preen gland waxes that have single-methyl substitutions at up to four separate positions in the carbon chain (Smith (12)). These compounds are substituted with two methyl groups at the alpha- or beta-positions, resulting in a quaternary carbon. The recalcitrance of these BCFAs was attributed to the presence of quarternary carbons which rendered cleavage by beta-oxidation impossible.

4. CONCLUSIONS

The bacteria and their roles in the anaerobiosis of 2-EHA were identified. The fate of 2-EHA in the anaerobic ecosystem was elucidated and beta-oxidation was proposed as the acidogenic mechanism. The consortium of acid-degrading *Syntrophomonas spp.*, H_2-utilizing *Methanococcus spp.* and acetoclastic *Methanothrix spp.* were found to degrade BCFAs with tertiary carbon but not those with quaternary carbon at the alpha or beta position.

REFERENCES

1. Masey, L.K., Sokatch, J. and Conrad, R.S., (1976) Bacteriol. Rev., 40, 42-54.
2. Ng, W.J., Yap, M.G.S. and Sivadas, M., (1989) Bio. Wastes, 29, 299-311.
3. Chua, H., Yap, M.G.S. and Ng, W.J., (1992) App. Biochem. Biotech., 34/35, 789-800.
4. Jimeno, A., Bermudez, J.J., Canovas-Diaz, M., Manjon, A. and Iborra, J.L., (1990) Bio. Wastes, 34, 241-250.
5. Richardson, A.J., Hobson, P.N. and Campbell, G.P., (1987) Lett. App. Microbiol., 5, 119-121.
6. Drier, T.M. and Thurston, E.L., (1978) Scanning Electron Microscopy, 11, 843-848.
7. Huser, B.A., Wuhrmann, K. and Zehnder, A.J.B., (1982) Arch. Microbiol., 132, 1-9.
8. Zehnder, A.J.B., Huser, B.A., Brock, T.D. and Wuhrmann, K., (1980) Arch. Microbiol., 125, 1-11.
9. McInerney, M.J., Bryant, M.P. and Pfennig, N., (1979) Arch. Microbiol., 122, 129-135.
10. McInerney, M.J., Bryant, M.P., Hespell, R.B. and Costerton, J.W., (1981) App. Env. Microbiol. 41(4), 1029-1039.
11. Jones, W.J., Nagle, D.P. and Whitman, W.B., (1987) Microbiol. Rev., 51(1), 135-177.
12. Smith, C. R., (1970) Topics in Lipid Chemistry, (Gunstone, F.D. (ed.)), Logos Press Limited, London, 277-368.

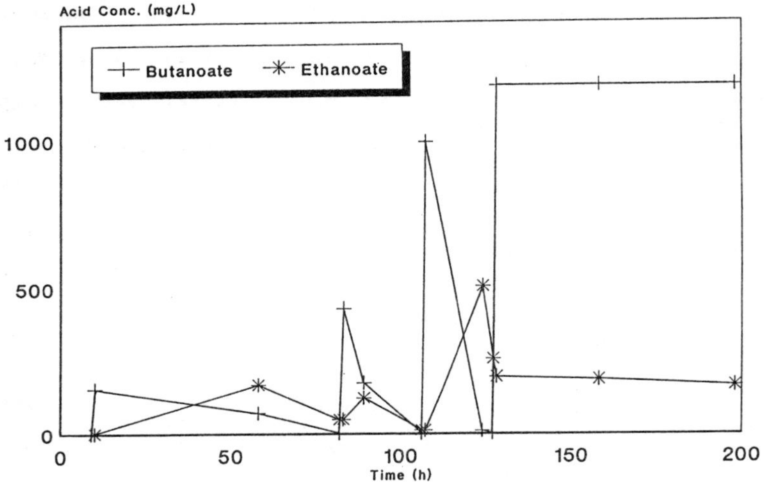

Figure 1 Acid Concentration Profiles in Butanoic-Acid Enrichment Culture

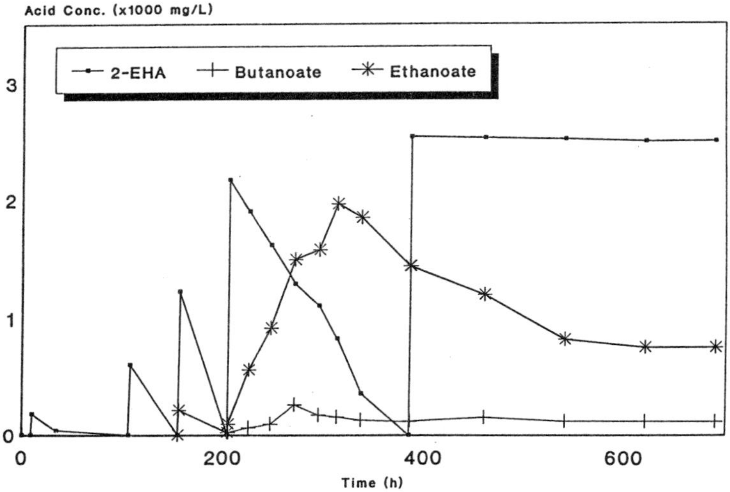

Figure 2 Acid Concentration Profiles in 2-EHA Enrichment Culture

Beta-oxidation :

Cleavage / Cleavage

$$C_4H_9-CH_2(C_2H_5)COOH \xrightarrow{2H_2O} 2C_2H_5-CH_2COOH + 2H_2 \xrightarrow{4H_2O} 4CH_3COOH + 4H_2$$

Decarboxylation :

$$4CH_3COOH \longrightarrow 4CH_4 + 4CO_2$$

Reduction :

$$6H_2 + 1.5CO_2 \longrightarrow 1.5CH_4 + 3H_2O$$

Overall reaction :

$$2\text{-EHA} + 3H_2O \longrightarrow 5.5CH_4 + 2.5CO_2$$

Figure 3 Mechanism of 2-EHA Degradation

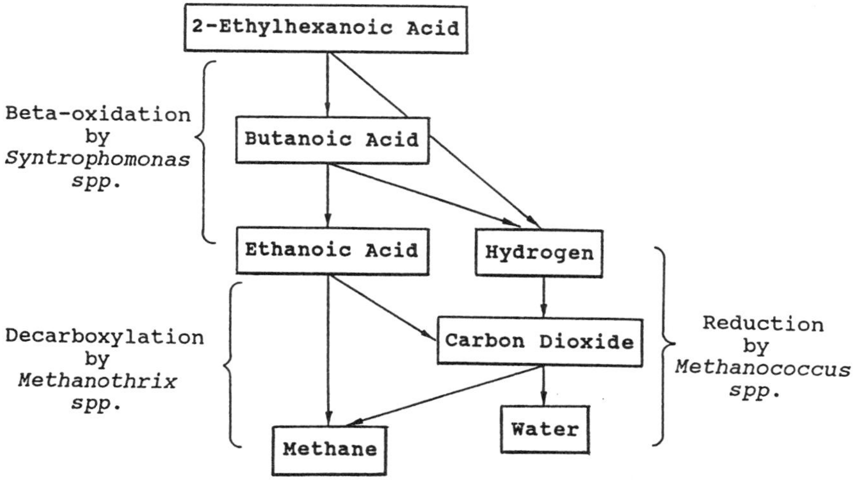

Figure 4 Fate of 2-EHA in Anaerobiosis

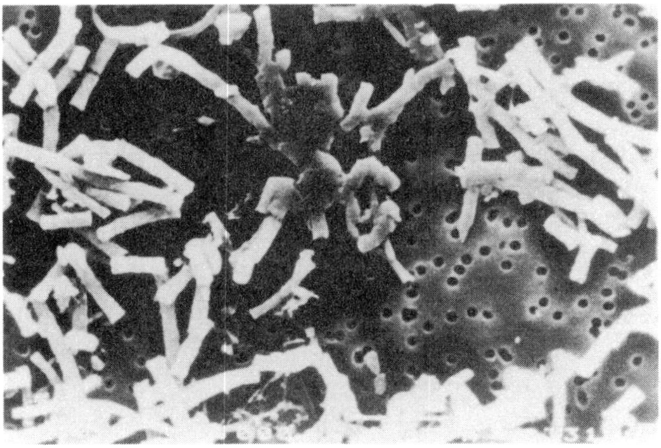

Plate 1 Ethanoic-Acid Enrichment Culture

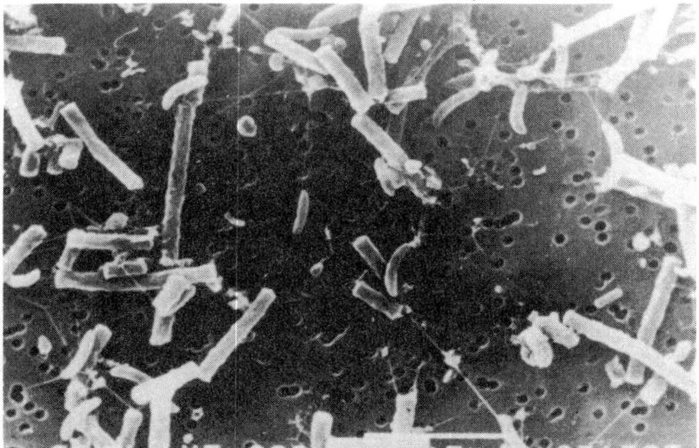

Plate 2 Butanoic-Acid Enrichment Culture

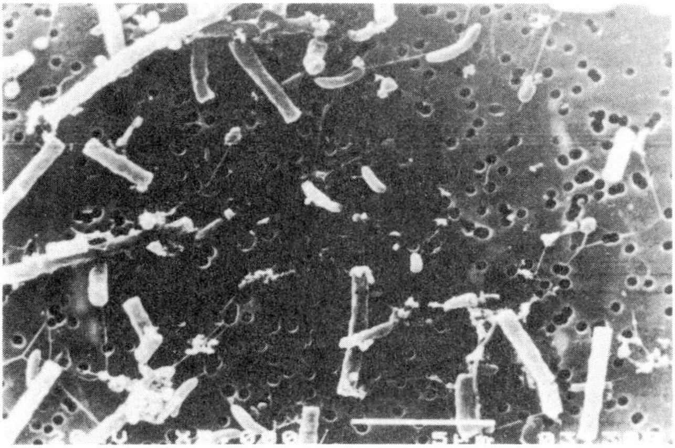

Plate 3 2-EHA Enrichment Culture

PRODUCTION OF ADRIAMYCIN AND OXYTETRACYCLINE BY GENETIC ENGINEERED *Streptomyces* WITH PALM OIL AND PALM KERNEL OIL AS CARBON SOURCES

Ho C.C., Melor I., Ong L.M., Sarimah A., Cheong M.P.K., Lee S.K., Yap C.C., and Tan E.L.
Department of Genetics and Cellular Biology, University of Malaya, Kuala Lumpur, Malaysia.

Palm oil was suitable for adriamycin and oxytetracycline production. *S. peucetius* var. *caesius* (H3502) could not grow on fatty acids but two oleic acid utilizers (H6310, H6360) with poor adriamycin production were isolated. Transformants of H6310 with *dnrR1* and *dnrR2* regulatory segments did not grow on oleic acid. The *dnrR2* transformants produced more adriamycin than H6310. The specific activities of fatty acy-CoA synthetase isocitrate lyase and malate synthase were higher in H6310 as compared to H3502 and its transformants. The results suggested that the products of *dnrR1* and *dnrR2* positively activate adriamycin biosynthesis but repress the genes for β-oxidation and glyoxylate pathway.

INTRODUCTION

The focus of this paper is to investigate the utilization of palm oil (mainly with C-16 and C-18 fatty acids), palm kernel oil (mainly with C-12 and C-14 fatty acids) their various fractions, their component fatty acids and glycerol as the main carbon sources for the clinically valuable anti-cancer antibiotics, adriamycin and daunomycin produced by *Streptomyces peucetius* var. *caesius*. Strain improvement for *S. peucetius* var. *caesius* was through the creation of transformants with introduced positive regulatory *dnrR1* and *dnrR2* DNA segments from *S. peucetius* (Stutzman-Engwall **et al.**, (1)) carried on multicopy pIJ702 plasmid. Two mutants (H6310, H6360) with ability to grow well on oleic acid and other long-chain fatty acids were isolated as the parent adriamycin producer (H3502) is unable to grow on fatty acids. The characteristics of H6310 and its transformants with *dnrR1* (H6322) and *dnrR2* (H6323) supported the hypothesis that while these regulatory segments particularly *dnrR2* stimulate the production of adriamycin, they also inhibit the utilization of fatty acids at the β-oxidation and glyoxylate pathways.

Similar work was carried out for the production of the industrially important anti-bacterial antibiotic, oxytetracycline produced by *Streptomyces rimosus* (H6336). In contrast to *S. peucetius* var. *caesius* (H3502), *S. rimosus* (H6336) was able to utilize well both palm oil, palm kernel oil and their component fatty acids (except lauric acid) for oxytetracycline production.

MATERIALS AND METHODS

Adriamycin Production

The parent strain used was H3502 and the basic medium was the AD1 medium (Ho and Chye, (2)) with substitution of sucrose by appropriate lipid or fatty acid. Adriamycin and daunomycin were isolated from the mycelia and partially purified following the procedure of Arcamone **et al.** (3). Identification and quantitation of adriamycin and its precursors was by reverse phase HPLC using a Waters C18 μBondapak P/N 27324 column with a mobile phase of acetonitrile and deionised water + 0.1% trifluroacetic acid (TFA) in a ratio of 25:75 at a flow rate of 2.5ml min^{-1} in the initial phase. The ratio was increased to 50:50 after 9 minutes and then returned to 25:75 after 14 minutes. Detection was at 254nm with 0.1 auf sensitivity.

Oxytetracycline Production

The strain used was M4018, Laboratory stock no. H6336, (Butler **et al.** (4)) and the basic medium was also the AD1 medium. Oxytetracycline was released from the mycelia and culture fluid by acidification (pH 2.5) followed by freeze drying of the extract. Identification and quantitation of oxytetracycline was done by HPLC under the following conditions: column, Inertsil ODS-2; mobile phase, methanol (166.7ml): acetonitrile (250ml): 0.01M oxalic acid (1000ml); flow rate, 1 ml min^{-1}; detector, at 280nm with 0.1 auf sensitivity. Samples were prepared by dissolving 10mg of the freeze dried crude antibiotic extract in 2.5ml of 2% HCl in methanol and diluting 0.1ml of the aliquot in 4.9 ml Na_2EDTA Mc-Ilvaine buffer (pH 4.0). For both adriamycin and oxytetracycline, fermentation were carried in shake flasks (220rpm) incubated at 28°C.

Construction of Transformants With Extra Copies of Regulatory Genes

The transformants were constructed through PEG-aided protoplast transformation with the following regulatory genes carried on multicopy plasmid (particularly pIJ702):

1. DnrR1 (a 2.0kb insert containing the regulatory genes dnrI) and
2. DnrR2 (a 1.9kb insert containing the regulatory gene dnrN) isolated from *S. peucetius*. DnrR1 and DnrR2 were kindly provided by C.R. Hutchinson, University of Wisconsin, Madison.
3. afsR (pIJ702-AP22, a 2.1kb, coding the DNA binding carboxyl end of the afs-R protein) isolated from *S. coelicolor*, a gift of S. Horinouchi, University of Tokyo.
4. abaA (a 2.7kb insert carried on pIJ486) isolated from *S. coelicolor*, a gift of M.A. Fernandez-Moreno, National Centre of Biotechnology, Madrid.

The transformants were selected through thiostrepton resistance, whose gene has been inserted into the plasmid.

RESULTS

With regard to the utilization of palm oil and palm kernel oil for production of adriamycin and daunomycin, the liquid fraction of palm oil, NBD (neutralized, bleached, deodorized) palm olein was shown earlier to be the best palm oil fraction for the production of these antibiotics. The results of repeated experiments (Table 1) showed that these antibiotics were produced in smaller amount with NB palm kernel olein than NBD palm olein. This trend was seen in spite of variability in the yield of these antibiotics between repeated experiments. As for the other fractions of palm kernel oil, the solid palm kernel stearin was slightly superior to palm kernel oil and palm kernel olein particularly for the production of daunomycin.

As the original strain, H3502 produced very small amounts of adriamycin and daunomycin, genetic engineered strains with better production were constructed. Two transformants of H3502 namely H6350 (with dnrR1 DNA segment) and H6357 (with dnrR2 DNA segment) with the regulatory segments carried on multicopy pIJ702 plasmid were constructed. Presumably these transformants will contain extra copies of the regulatory genes contained in the dnrR1 and dnrR2 segments leading to increased transcription of the structural genes for the biosynthesis of adriamycin. In shake flask experiments using NBD palm olein (1%) replacing sucrose as the carbon source, H6357 produced more adriamycin while H6350 produced more ε-rhodomycinone than its parent, H3502. (Table 2).

A striking feature is that while *S. peucetius* var. *caesius* (H3502) is able to grow well on palm oil, its olein and stearin fractions and pure triglycerides but unable to grow on its free fatty acids namely lauric, myristic, oleic, linoleic, linolenic, stearic and palmitic (with trace of growth). In an intensive mutagenesis programme to isolate mutants that will grow well on fatty acids with good production of adriamycin, two oleic acid utilizing mutants (H6310, H6360) were isolated. The first mutant, H6310 white in colour with massive sporulation was isolated after NTG mutagenesis and produced a fairly large amount of aklanonic acid, the first intermediate of adriamycin biosynthesis together with very small amounts of adriamycin and daunomycin (Ho **et al.** (5)). The second mutant, H6360 was isolated after EMS mutagenesis, had orange yellow mycelia instead of the red colour of its parent (H3502). H6360 in AD1 medium supplemented with NBD palm olein (1%) replacing sucrose it produced lesser amount of adriamycin, daunomycin and their precursor, ε-rhodomycinone as compared to its parent. The mechanism involved in the utilization of lipid and fatty acid for adriamycin production was therefore investigated in the following manner.

Extracellular Lipase From *S. peucetius* var. *caesius*

The extracellular lipase from H3502 and its oleic acid utilizing mutant (H6310) when grown in AD1 medium with sucrose replaced by NBD palm olein (2%) was studied. For both strains, the degradation of triglycerides was parallel with rise of free fatty acids and glycerol with H6310 showing lesser accumulation of free fatty acids and glycerol as compared to H3502.

This was confirmed by thin layer chromatography showing massive accumulation of free fatty acids in H3502 only (Figure 1). The lipases from both of these have been partially purified to 50 and 30 fold for H3502 and H6310 respectively. The purified enzymes were then characterized. Lipolysis profile of NBD palm olein from H3502 was similar to H6310 and required 8 hours for complete hydrolysis. Optimum conditions for lipolysis of NBD palm olein were 50°C, pH 8 for H3502 lipase and 40°C, pH 7 for H6310 lipase. The lipase from both strains were non regio-specific producing 1,2 (2,3)-diglycerides, 1,3-diglycerides and barely detectable monoglycerides.

Transformants (dnrR1 and dnrR2) of H6310 and Their Abilities to Grow On Fatty Acid and Adriamycin Production

Transformants of H6310 with dnrR1 (H6322) and dnrR2 (H6323) were constructed and their abilities to grow on fatty acids and adriamycin production were examined (Table 3). Interestingly, the transformants like H3502 were unable to grow on oleic acid. On triglyceride (palm olein) medium, the dnrR2 transformant showed a small increase in the production of adriamycin and daunomycin as compared to H6310 but still less than the original adriamycin producer (H3502). When grown on palmitic acid (1%) H3502 produced a small amount of adriamycin (30µg) and daunomycin (100µg) per gram dry cell weight while H6310 and its transformants were unable to do so (Table 4).

These preliminary results suggest that extra copies of the positive regulatory DNA segments (dnrR1 and especially dnrR2) stimulated adriamycin biosynthesis while at the same time also act to repress the transcription of the genes for β-oxidation of long chain fatty acids to acetyl CoA and the glyoxylate pathway. This hypothesis was confirmed by the finding that the specific activities of long-chain fatty acyl Co A synthetase (Barr-Tana, Rose and Shapiro, (6)), isocitrate lyase and malate synthase (Dixon and Kornberg, (7)) were higher in the oleic utilizer (H6310) as compared to its parent (H3502) and its transformants with dnrR1 and dnrR2 segments (Table 5). Similar results were also obtained for the second oleic acid utilizer, H6360 and its transformants.

Oxytetracycline Production

Streptomyces rimosus was able to utilize the lipids and fatty acids both for growth and oxytetracycline production. The lipids were better than the fatty acids and glycerol for oxytetracycline production. Amongst the oils, palm stearin gave the highest titre of oxytetracycline. Palm kernel oil and its fractions, palm kernel olein and palm kernel stearin surprisingly produced substantial amounts of oxytetracycline amounting to almost as good as crude palm oil and its fractions (Table 6).

Oleic acid was the best among the fatty acids for oxytetracycline production giving titres almost as good as the oils. Other fatty acids namely myristic, palmitic and stearic, although not as good as the oils and glycerol were still capable of producing considerable amounts of growth and oxytetracycline. Lauric acid was inhibitory to growth with no production of oxytetracycline. The triglycerides were definitely superior to glucose and glycerol while the fatty acids were either better than, or at least equal to glucose for growth and oxytetracycline production.

In efforts to create genetic engineered strains with higher oxytetracycline production, transformants of *S. rimosus* (H6336) with introduced heterologous regulatory genes, afsR (Horinouchi **et al.** (8)) and abaA (Fernandez-Moreno **et al.** (9)) isolated from *S. coelicolor* and carried on multicopy plasmids were kindly constructed by Kantilal H. K. of this laboratory. These regulatory genes originally stimulated the synthesis of actinorhodin (a polyketide antibiotic) in *S. coelicolor*. Surprisingly afsR seems to completely shut off, while abaA decreased by half oxytetracycline production (Table 7).

DISCUSSION

Palm oil but not palm kernel oil is suitable and superior to glucose for growth and adriamycin production in *S. peucetius* var. *caesius*, and its improved genetic engineered transformant with multicopies of dnrR2 regulatory DNA segment. This segment is known to contain the regulatory gene, dnrN for daunomycin biosynthesis (Hutchinson **et al.**, (10)). The mutation that results in strain H6310 and H6360 growing well on oleic acid is rather similar to the regulatory gene, fadR in *Escherichia coli* which represses the fad regulon which include genes for β-oxidation and fatty acid transport but also positively activates a fatty acid biosynthetic gene, fabA (DiRusso, Heimert and Metzer, (11)). The effect of dnrR1 and dnrR2 to inhibit fatty acid utilization while at the same time stimulating the biosynthesis of adriamycin suggests the interesting possibility that the mutated gene in H6310 and H6360, and the regulatory gene in dnrR1 and dnrR2 in S. peucetius may be homologous to the fadR gene. To our knowledge, this is the first demonstration of a common regulation for fatty acid utilization and polyketide (adriamycin) biosynthesis.

Palm oil and palm kernel oil and their fatty acids (except lauric acid) are very suitable for oxytetracycline production. Already it is well known that vegetable oils are suitable for industrial production of oxytetracycline (Abou-Zeid and Baeshin, (12)) The heterologous regulatory genes, afsR (Ho, Nissom and Krishna (13)) and abaA (Ho C. C., Ong T. F. and Krishna G., unpublished result) have been shown in this laboratory to stimulate adriamycin production in S. peucetius var. caesius but inhibit oxytetracycline production in S. rimosus. This unexpected result in S. rimosus is similar to the report of Butler **et al.** (4) that a cloned DNA fragment in multicopy vectors often switched off oxytetracycline production in this organism.

Further investigations into the molecular mechanism of the linkage between fatty acid metabolism (degradation and biosynthesis) and polyketide synthesis in Streptomyces and its control by regulatory genes, particularly through signal transduction by protein phosphorylation and specific DNA binding for gene transcription (Horinouchi and Beppu, (14)) will be of fundamental and applied importance.

REFERENCES

1. Stutzman-Engwall K. J., Otten S. L. and Hutchinson C.R. 1992. Regulation of secondary metabolism in Streptomyces spp. and overproduction of daunorubicin in Streptomyces peucetius. **J. of Bacteriology** 174 144-154.
2. Ho C.C. and Chye M. L., 1985. Construction of a genetic map of chromosomal auxotrophic markers in Streptomyces peucetius var. caesius. **J. Gen Appl. Microbiol.** 31 231-241.
3. Arcamone F., Cassinelli G., Fantini G., Grein A., Orezzi P., Pol C., and Spalla C. 1969. Adriamycin, 14-hydroxydaunomycin, a new antitumor antibiotic from S. peucetius var. caesius **Biotech and Bioeng.** 11 1101-1110.
4. Butler M. J., Friend E. J., Hunter I. S. Kaczmarek F. S., Sugden D. A. and Warren M., 1989. Molecular cloning of resistance genes and architecture of a linked gene cluster involved in biosynthesis of oxytetracycline by Streptomyces rimosus **Mol. Gen. Genet.** 215 231-238.
5. Ho C.C., Lee S.K., Benjamin D.G., Ong L.M., Cheong M.P.K., Krishna G. and Tan E.L., 1991. A common regulatory genetic control (dnrR1 and dnrR2) for adriamycin biosynthesis and fatty acids utilization for growth in Streptomyces peucetius var. caesius **Abstract 26, 3rd. Seminar National Biotecnology Programme**, Kuantan, Pahang, Malaysia.
6. Barr-Tana J., Rose G. and Shapiro B., 1971. The purification and properties of microsomal palmitoyl-Coenzyme A synthetase **Biochem. J.** 122 353-362.
7. Dixon G. H. and Kornberg H. L., 1959. Assay methods for key enzymes of the glyoxylate cycle **Biochem. J.** 72 3p.
8. Horinouchi S., Kito M., Nishiyama M., Furaya K., Hong S.K., Miyake K. and Beppu T., 1990. Primary structure of afsR, a global regulatory

protein for secondary metabolite formation in Streptomyces coelicolor A3(2) **Gene** 95 49-56.
9. Fernandez-Moreno M. A., Martin-Triana A. J., Martinez E., Niemi J., Kieser H. M., Hopwood D. A. and Malpartida F., 1992. abaA, a new pleiotropic regulatory locus for antibiotic production in Streptomyces coelicolor **J. Bacteriology** 174 2958-2967.
10. Hutchinson C. R., Decker H., Madduri K., Otten S. L. and Tang L. 1993 Genetic control of polyketide biosynthesis in the genus Streptomyces **Antonie van Leeuwenhoek** 64 165-176.
11. DiRusso C. C., Heimert T. L. and Metzer A. K., 1992. Characterization of FadR, a global transcriptional regulator of fatty acid metabolism in Escherichia coli **J. Biol. Chem.** 267 8685-8691.
12. Abou-Zeid A. A. and Baeshin N. A., 1992. Utilization of date-seed lipid and hydrolysate in the fermentative formation of oxytetracycline by Streptomyces rimosus **Bioresource Tech.** 41 41-43.
13. Ho C. C., Nissom P. M. and Krishna G., 1993. Stimulation of adriamycin biosynthesis in Streptomyces peucetius var. caesius by multiple copies of afsR, a global regulatory gene of Streptomyces coelicolor **Abstract 4th. Scientific Meeting, Malaysian Society for Molecular and Cellular Biology** 19-20 May 1993, Genting Highlands, Pahang, 40-41.
14. Horinouchi S. and Beppu T., 1992. Regulation of secondary metabolism and cell differentiation in Streptomyces: A factor as a microbial hormone and the afsR protein as a component regulatory system **Gene** 115 167-192.

	NBD palm olein	NB palm kernel olein
Adriamycin	7.2 ± 5.53 (7)	1.9 ± 2.61 (7)
Daunomycin	23.1 ± 16.8 (7)	9.2 ± 8.94 (7)

TABLE 1 Antibiotic production (μg/g dry cell weight) from 72 hour culture of H3502 grown in modified AD-1 medium with lipid (2%). Number in parenthesis refers to number of experiments. Mean ± SD.

	Adriamycin	Daunomycin	ε-rhodomycinone
H3502	45.3 ± 35.09 (5)	29.5 ± 19.73 (5)	10.8 ± 6.98 (5)
H6350(dnrR1)	3.9 ± 1.73 (3)	24.8 ± 15.45 (3)	77.1 ± 29.82 (3)
H6357(dnrR2)	221.2 ± 142.68 (4)	113.0 ± 146.39 (4)	9.6 ± 6.33 (4)

TABLE 2 Production of anthracyclines and ε-rhodomycinone (μg/g dry cell weight) by 72 hour cultures on NBD palm olein (1%) with 5μg/ml thiostrepton added for transformants. Figure in parenthesis refers to number of experiments. Mean ± SD.

		H3502	H6310	H6322(dnrR1)	H6323(dnrR2)
A	**Growth**				
	Palmitic acid	2.8[a]	2.5	0	0
	Oleic acid	0	3.5	0	0
	NBD Palm olein	2.3	2.3	2.8	2.6
B	**Adriamycin and precursors production on NBD palm olein**				
	Adriamycin	210	0	0	11
	Daunomycin	156	9	1.25	29
	ε-rhodomycinone	1.4	0	0	0
	Aklavinone	0	9	1.97	1.97
	Aklanonic acid	0	21	0	0

a - highly contaminated with palmitic acid

TABLE 3 Growth (mg/ml) and production of adriamycin and precursors (μg/gm cell dry weight) at 96 hour for H6310 and its transformants. Medium used was AD-1 with substitution of sucrose by lipid or fatty acid (1%)

	H3502	H6310	H6322(dnrR1)	H6323(dnrR2)
Palmitic acid (1%)				
Adriamycin	30	0	NG	NG
Daunomycin	100	0	NG	NG
ε-rhodomycinone	31	0	NG	NG
Aklavinone	490	2	NG	NG
Aklanonic acid	0	12	NG	NG
Oleic acid (1%)				
Adriamycin	NG	0	NG	NG
Daunomycin	NG	0	NG	NG
ε-rhodomycinone	NG	0	NG	NG
Aklavinone	NG	8	NG	NG
Aklanonic acid	NG	0.24	NG	NG

0 - No production of antibiotics
NG - No growth

TABLE 4 Growth(mg/ml) and production of adriamycin and precursors (μg/gm dry cell weight) by 96 hour cultures in fatty acid medium

Enzyme	H3502	H6310	H6322(<u>dnrR1</u>)	H6323(<u>dnrR2</u>)
I Long-chain fatty acyl-CoA synthetase (nmol/hr/mg protein) assayed as				
a Palmitoyl-CoA synthetase	148	463	157	221
b Oleoyl-CoA synthetase	41	345	40	23
II Isocitrate lyase (nmol/min/mg protein)	1	2.1	0.8	0.95
III Malate synthase (nmol/min/mg protein)	400	526	202	147

TABLE 5 Specific activities of fatty long-chain acyl-CoA synthetase, isocitrate lyase and malate synthase in 96 hour cultures grown in AD-1 medium in shake flasks

CARBON SOURCE (1%)	% Purity	Final pH	Cell yield (mg/ml)	µg OTC per ml medium
CRUDE PALM OIL		7.8	4.50	328.9
NBD PALM OLEIN		7.8	6.74	271.8
NBD PALM STEARIN		7.9	6.00	395.3
NB PALM KERNEL OIL		7.7	6.48	284.9
NB PALM KERNEL OLEIN		7.8	6.33	317.1
NB PALM KERNEL STEARIN		7.8	5.51	369.2
LAURIC ACID	99-100	6.6	1.37	0
MYRISTIC ACID	98	7.5	3.12	134.5
PALMITIC ACID	95	7.4	5.28	111.2
STEARIC ACID	90	7.3	8.16	124.0
OLEIC ACID	92	8.1	5.69	300.3
GLYCEROL		8.0	2.48	223.6
GLUCOSE		7.9	2.49	120.7

Initial pH of medium was adjusted to 6.8 after addition of lipids and fatty acids.
The oils were obtained from Unitata Plantations, Teluk Intan, Perak.
NBD = Neutralised, Bleached, and Deodourised
NB = Neutralised and Bleached

TABLE 6 Quantitative analyses of the fermentation cultures (96 hour) of *Streptomyces rimosus* H6336 (based on an average of 5 experiments with exceptions for myristic, palmitic and oleic acids [4] and glycerol [3]).

CARBON SOURCE (1%)	H6336				H6396(afsR)				H6412(abaA)			
	Final pH	Cell Yield mg/ml	µg OTC/ ml medium	µg OTC/ mg dry cells	Final pH	Cell Yield mg/ml	µg OTC/ ml medium	µg OTC/ mg dry cells	Final pH	Cell Yield mg/ml	µg OTC/ ml medium	µg OTC/ mg dry cells
NBD PALM OLEIN	7.9	8.21	342.9	41.5	7.5	7.38	0	0	7.8	7.20	139.0	22.8
OLEIC ACID (92% Pure)	8.3	5.08	293.0	57.7	7.9	6.33	0	0	7.9	6.75	179.0	26.8
GLUCOSE	7.4	2.48	144.1	58.7	7.9	4.72	0	0	8.5	4.11	101.2	25.1

All average values are based on 5 repeated fermentations except for H6336 (4) and H6396 (oleic acid) (4). Samples were prepared as in text with modifications; samples from transformants were prepared using 30mg of crude extract; for all samples, after diluting the crude extract in 2.5ml 2% HCl in MeOH, 1.0ml of the aliquot was diluted in 49ml Na_2EDTA Mc-Ilvaine buffer.

TABLE 7 Quantitative analyses of production cultures (96 hour) of *Streptomyces rimosus* H6336, H6396, and H6412 grown with 1% palm olein, oleic acid and glucose

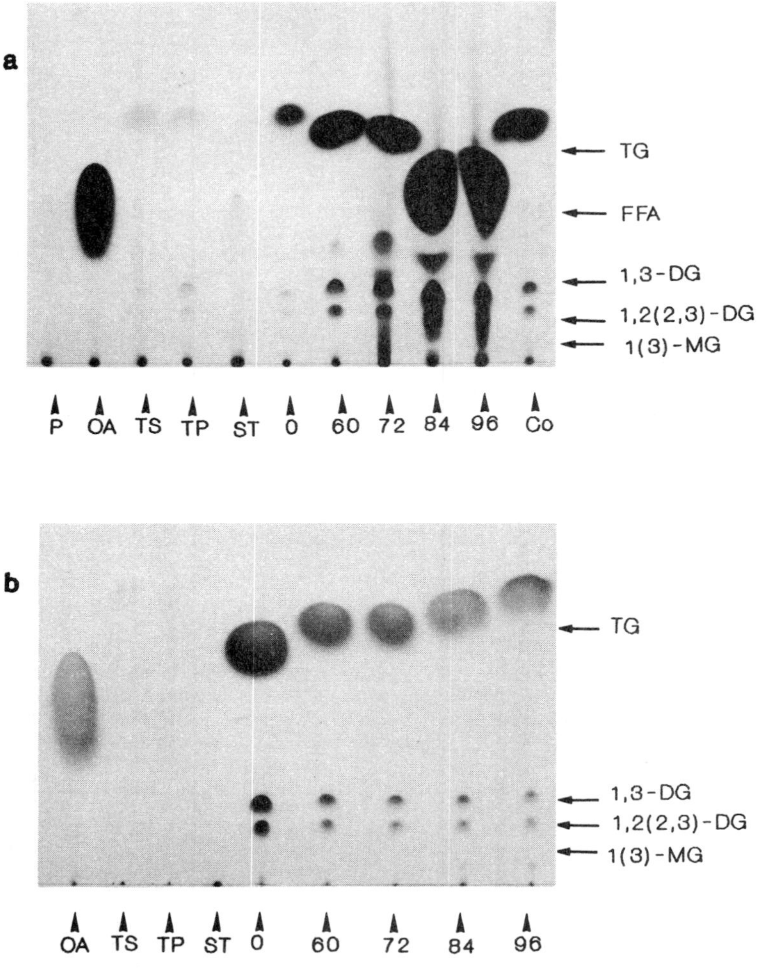

FIGURE 1 Thin layer chromatograms of palmitic acid (P), oleic acid (OA), tristearin (TS), tripalmitin (TP), stearic acid (ST) and lipid extracts from production cultures of **a** H3502 and **b** H6310 at incubation times of 0 h (0), 60 h (60), 72 h (72), 84 h (84) and 96 h (96).
Monoglycerides (MG), diglycerides (DG), free fatty acids (FFA), triglycerides (TG) and NBD palm olein positive control (Co).

PRODUCTION OF RECOMBINANT PROTEINS BY HIGH PRODUCTIVITY MAMMALIAN CELL FERMENTATIONS

P P Gray and K Jirasripongpun
Bioengineering Centre, Department of Biotechnology, University of NSW, Sydney 2052, Australia

Growth of a recombinant CHO cell line expressing r-hGH on non-porous (Cytodex 2) and porous (Cultisphere-G) microcarriers was assessed in a continuous perfusion bioreactor. Increased concentrations of the microcarriers, coupled with a continuous feed of medium containing 2.5% FCS, were used to obtain high final cell densities in order to increase the volumetric productivities which could be obtained for such perfusion cultures. Final cell concentrations increased with increasing concentrations of Cytodex-2 up to a final value of 3×10^7 cells/ml. Increases in Cultisphere-G concentration did not result in any further increases in cell concentration above the 8.2×10^6 cells/ml obtained under control conditions.

1. INTRODUCTION

The last few years has seen a rapid increase in the use of mammalian cell cultures for the production of biopharmaceuticals. Mammalian cells are being used increasingly as hosts for the production of recombinant DNA derived proteins and antibodies. Many of the difficulties associated with the large-scale growth and product formation from such cells are being overcome as development work progresses on reactors and growth conditions. From a process viewpoint, one of the inherent problems with mammalian cell fermentations is the low cell densities reached in such cultures, maximum cell densities usually being of the order of 2-5 x 10^6 cells/ml. This low cell density is a major factor in limiting the volumetric productivity which can be obtained from mammalian cell fermentations. Such cultures utilise very complex media which have usually been developed under static batch culture conditions, and it is often difficult to determine the factors which limit the final cell concentrations obtained. In our research we have been interested in maximizing the final cell concentration in order to produce concommitant increases in the volumetric productivities which can be obtained for recombinant protein being expressed by mammalian cell cultures.

In this paper a comparison has been made between the use of porous and non-porous microcarriers as supports for the growth and subsequent product formation by a recombinant Chinese Hamster Ovary (CHO) cell line expressing recombinant human growth hormone under the control of the metallothionein promoter.

2. MATERIALS AND METHODS

The CHO cell line grown in the study (CB515) was constructed so as to express high levels of recombinant hGH without the need for high levels of gene amplification, Friedman *et al* (1).

The cells were grown on microcarriers in a stirred reactor with an operating volume of approximately one litre. Microcarriers and their attached cells were retained in the bioreactor by stainless steel screens of 100μm pore size. Cells were grown on a DMEM:COONS F12 medium containing 10% FCS. In order to obtain cell densities which were higher than those which can usually be obtained in simple batch culture (of the order of 2-5 x 10^6 cells/ml), the final stages of the growth phase were run as continuous perfusion cultures, a continuous feed of medium containing 2.5% FCS being started once the residual glucose level fell to 0.5 g/l. At the end of the growth phase, the medium was replaced with a protein free production medium (DMEM:COONS F12) containing zinc to induce expression of the hGH. All other procedures and assays were as previously reported (Gray *et al* (2)).

The microcarriers used in the study were Cytodex-2, a dextran based microcarrier obtained from Pharmacia Fine Chemicals and Cultisphere-G, a gelatin based microcarrier obtained from Percell Biolytica.

The physical characteristics of the two microcarriers are shown in Table 1.

Table 1: Properties of Microcarriers.

	Cytodex-2	Cultisphere-G
Density (g/L)	1.04	1.04
Size (mm)	114-198	170-270
Approx area (cm^2/g dry weight)	5500	Unknown
Approx no (beads/g dry weight)	5.8 x 10^6	2.3 x 10^6
Matrix	Dextran	Gelatin

3. RESULTS AND DISCUSSION

Cytodex-2 microcarriers are solid and provide only the outer surface of the microcarrier for cell attachment and growth. By comparison, Cultisphere-G are porous microcarriers which theoretically provide both inner surfaces as well as outer surfaces for cell attachment and growth. Other differences between the two types of microcarrier are listed in Table 1 above. Cultisphere-G would appear to have the advantage of a higher surface area to unit volume ratio and provide an environment which offers the cells greater protection from shear stress. A comparison was carried out using the two microcarriers in a stirred one litre reactor.

Cell inoculum for the experiments using Cultisphere-G was produced by cultivating the cells in 10% FCS medium in a 500 ml spinner flask at 25-30 rpm for one day before transferring the cells to the one litre stirred fermentor. This procedure was required due to the slower rate of attachment of the cells to the Cultisphere-G (Loopy and Griffiths (3)). For the cultivation of Cytodex-2 the culture inoculum was transferred direct from a 490 cm^2 roller bottle. After transferring the inoculum to the fermentor both cultures were cultivated in the same manner, viz. batch mode in 10% FCS until the glucose was limiting, then a feed of 2.5% FCS medium until cell concentration plateaued when the medium was changed to a production protein-free medium. The feed rate of the production medium was maintained at 0.3 vol / 10^6 cells/day. Oxygen in the culture was controlled at 60% of air saturation.

The results are shown in Table 2. At all Cultisphere-G concentrations the maximum cell density reached was approximately 8 x 10^6 cells/ml. The nutrients in the cultures were found to be not limiting (glucose was between 0.5-1.0 g/l in all cases, while waste products were low; lactic acid and ammonium were around 0.7-0.85 g/l and 1 mmol respectively). Under these conditions the cell density in the culture should have increased with the Cultisphere-G concentration; the results shown in Table 2 show that the final cell density was independent of the final concentration of microcarrier. Similar experiments were carried out using the Cytodex-2 microcarrier: for these microcarriers the control condition used 3 g/l of microcarriers at which concentration it was usual to obtain of the order of 3 x 10^6 cells/ml. Table 3 shows results obtained when the concentration of microcarriers was increased to 8 and 12 g/l. It can be seen that increasing the microcarrier concentration in the perfusion bioreactor resulted in an increase in cell density from 1.8 x 10^6 cells/ml at 8 g/l to 3 x 10^7 cells/ml at 12 g/l. Comparison of Tables 2 and 3 show that the suspended cell densities were greater, after the changeover to production media, with the Cultisphere-G fermentations than with the Cytodex-2. The free cell densities in Cultisphere-G cultures were about 12-23% of the attached cell density compared with free cell densities obtained from Cytodex-2 cultures of 5-10% of the attached cell density.

Table 2: Growth of CB515 on Cultisphere-G in a 1 litre perfusion bioreactor

Microcarrier (g/L)	Attached cells (cells/ml)	Suspension cells (cells/ml)	Sus/Att cell (%)
3	8.24 x 10^6	1 x 10^6	12.14
5	8.46 x 10^6	1.3 x 10^6	15.37
7	8.40 x 10^6	2 x 10^6	23.81

In Table 4, Cytodex-2 and Cultisphere-G cultures were compared under similar conditions. The microcarrier concentration in each case was 5 g/l in a one litre stirred reactor; 60% air saturation dissolved oxygen level; medium perfusion rate of 0.3 vol x 10^6 cells/day or 2.5% FCS followed by changeover to the production medium.

Table 3: Growth of CB515 on Cytodex-2 in a 1 litre perfusion bioreactor

Microcarrier (g/L)	Att cell (cells/ml)	Sus cell (cells/ml)	Sus viab (%)
8	1.8 x 10^7	1 x 10^6	90
12	3.0 x 10^7	3 x 10^6	90

As previously shown, the attached cell density was higher for the Cytodex-2 culture at 1 x 10^7 cells/ml compared with 8 x 10^6 cells/ml for the Cultisphere-G. The calculated value of the growth rate was also higher for the Cytodex-2 than for the Cultisphere-G culture. Furthermore, cell attachment during cultivation with the porous microcarrier is difficult, even using the recommended procedures described previously. The Cytodex-2 cultures could be maintained in production phase for longer periods, often of the order of 20-30 days. By contrast it was observed that there was a rapid decrease in cell numbers on changing production medium in the Cultisphere-G cultures. This rapid decrease in cell numbers could have been one of the factors causing the increased specific productivity (q_{hGH}) in the Cultisphere-G cultures over Cytodex-2 as shown in Table 4. It has previously been observed that cultures with declining cell numbers often tended to exhibit increased specific productivity.

In Table 4 the metabolism of both cultures is compared. Glutamine uptake rates were similar, while glucose uptake rates and lactate production rates in the Cultisphere-G cultures were about two times higher than in the Cytodex-2 culture. It appears that the porous microcarrier environment enhanced the utilization of glucose resulting in high productivity of lactate. One possible explanation of this observation could be that the large diameter of the Cultisphere carriers of 170-270 mm could result in oxygen limitation inside the pores of the beads. Glacken *et al* (4) reported that oxygen limitation was caused at a diameter above 170 mm. It is possible that there may be oxygen limitations inside the porous microcarriers which limited the density of cells which could be supported per microcarrier.

Table 4: Comparison of metabolic parameters for CB515 cells growing on 5g/l of Cytodex-2 and Cultisphere-G in a 1 litre stirred bioreactor

Microcarrier	m (h^{-1})	Max cell (cells/ml)	hGH prod ($mg/10^6/$)	qglu	qlact ($mg/10^6/h$)	qglut	qammo ($mmole/10^6/h$)
Cytodex-2	0.030-0.090	1×10^7	0.69-0.83	16-19	16-20	20-30	23-25
Cultisphere-G	0.023-0.028	8×10^6	0.87-1.60	30-35	23-37	23-26	15-20

Although intuitively it was felt that the porous microcarriers would allow for even distributed growth of the cells throughout the internal volume of the microcarrier, it may be that the nature of the Cultisphere matrix was not particularly conducive for cell growth and the cells found difficulty in penetrating inside the beads in order to attach and grow. This factor could explain the numbers of suspended cells in the culture. Previous work had shown that cells grown inside the porous carriers tended to cluster in several regions of the beads and not disperse evenly throughout the internal pores of the microcarriers. Obviously concentration of cells in this fashion would lead to exacerbation of oxygen transfer problems and result in a microenvironment which would favour the concentration of waste products.

4. DISCUSSION

The technical literature accompanying the Cultisphere-G (Hyclone 'Cultisphere-G' microporous gelatin micro-carrier, *Technical Brochure*) reported final concentration of approximately 1×10^7 cells/ml for CHO cells growing on Cultisphere-G. With the cell line and conditions used in this study, it was not possible to get cell concentration greater than 8×10^6 cells/ml even when up to 7 g/l of the Cultisphere was used. By comparison, with the non-porous microcarrier, Cytodex-2, where cell growth was restricted to the surface of the microcarriers, cell densities of up to 3×10^7 cells/ml could be obtained at microcarrier concentrations of 12g/l. It had been hoped that the porous nature of the Cultisphere-G beads would allow even distributed growth of the CHO cells throughout the inner spaces of the beads. This did not appear to happen, and electron micrographs of the internal structures of the beads after growth showed that the cells tended to grow in a few 'clusters' inside the microcarriers. The size of the Cultisphere microcarriers, 170-270 mm could lead to oxygen and nutrient limitation inside the pores of the beads. It has been reported (Glacken *et al* (4)) that oxygen limitation occurs at a microcarrier diameter above 170mm. Experimental evidence has been given (Sutherland (5)) that there are steep gradients of oxygen near the periphery and very low concentrations beyond about 100µm from the surface. Diffusion limitations could explain why there might be a limitation on the number of cells which could be supported per microcarrier, but does not supply an adequate explanation as to why there was not a linear increase in total cell concentration with increasing microcarrier concentration. The number of cells in suspension with the Cultisphere-G was unusual, and it may be that the gelatin used for the microcarriers was not conducive to cell attachment, and this fact, coupled with the internal microenvironment of limiting nutrients and oxygen stimulated cell mobility to the liquid medium. It is interesting to note that the concentration of cells in suspension did increase with increasing concentration of Cultisphere-G (Table 2). Previous work by other workers in the group had indicated that non-porous microcarriers constructed of or coated with gelatin had problems of cell detachment under serum free conditions.

The aim of increased volumetric reactor productivity as a result of increased cell density was in fact demonstrated using the non-porous microcarriers, Cytodex-2. Using the methods described above, it was possible to obtain final cell densities of 5×10^7 cells/ml using 15g/l of the microcarriers. Such bioreactors could be run at throughputs of up to 12 volumes per day, under which conditions they were producing 0.78 grams per litre per day of the recombinant protein(Gray *et al* (6)). Such productivities are comparable with the values which can be obtained for prokaryotic cultures.

5. REFERENCES

1. Friedman, J.S., Cofer, C.C.L., Anderson, C.L., Kushner, J.A., Gray, P.P., Chapman, G.E., Stuart, M.C., Lazarus, L., Shine, J. and Kushner, P.J. (1989) Bio/Technology, 7, 359-362.
2. Gray,P.P.,Crowley,J.M. and Marsden,W.L. (1990). In 'Trends in Animal Cell Technology', Ed. H. Murakami, VCH Kodansha, Tokyo, 265-270.
3. Loopy, D. and Griffiths, J.B. (1990) Trends in Biotech, 8, 204-209.
4. Glacken, M.W. Fleischaker, R.W., Sinskey, A.J. (1983) Ann. N.Y. Acad. Sci. 413, 355-372.
5. Sutherland, R.M. (1988). Science, 240, 177-184.
6. Gray,P.P.,Jirasripongpun,K., and Gebert,C. (1992). in 'Harnessing Biotechnology for the 21st Century', Eds. M.R.Ladisch and A.Bose, American Chemical Society, Washington DC, 197-201.

EFFECTS OF ALKALI, CELLULASE AND CELLOBIASE ON THE PRODUCTION OF SUGARS FROM PALM WASTE FIBRE

M. Anis, K. Das* and N. Ismail, Division of Food Technology, School of Industrial Technology,Universiti Sains Malaysia,11800 Penang, MALAYSIA
* Corresponding Author

The yield of fermentable sugars by saccharification of oil palm waste fibres is found to be markedly higher within 48 h when the substrate is swelled by alkali treatments than that when it is simply boiled with alkali. The conversion was as high as 58% when swelling was applied, whereas, it was only 47% with the boiling process. Thus, an increase of more than 10% could be attained by swelling. After this comparison, swelling conditions were then optimised under which the conversion was found to increase to 76% resulting in further increase of 18%. A larger increase in conversion up to about 90% could be achieved within same 48 h when cellobiase was added along with cellulase during hydrolysis of treated palm waste, thereby, resulting an improved yield of 14%.This increase, due to supplementation of cellulase with cellobiase, can be explained for the conversion of cellulose to glucose by alleviating inhibition by cellobiose which is, in turn, converted to glucose by cellobiase.

1. INTRODUCTION

Solid palm wastes, either palm oil industry-related by-products like oil palm press fibres (PPF), palm bunch fibres, or plantation-related oil palm trunk are a major source of agro-waste resource materials in Malaysia for economic bioconversion to useful energy-intensive chemicals and biomass protein foods.

The major preliminary steps in such bioconversion systems are pretreatment and enzymic saccharification of these materials. Cellulose and hemicellulose which are important components of these residues are largely protected from attack by micro-organisms and their associated enzyme systems. The inaccessibility to attack them is mainly due to the association of these polysaccharides with lignin, which act as a barrier shielding the polysaccharides.

To effectively hydrolyse cellulosics with cellulases, pretreatment of these raw materials is necessary to make the cellulose more accessible to enzymic attack.Pretreatment methods, which disrupt the highly ordered cellulose structure and the lignin-carbohydrate complex, remove lignin and increase the surface area accesible to enzymes, promote hydrolysis, and increase the rate and extent of hydrolysis of cellulose as shown by Fan *et al* (1).One effective technique is the pretreatment of residues with cellulose-swelling or dissolving agents used by Ladisch *et al* (2) which reduce cellulose crystallinity and which may also disrupt the lignin seal in native cellulosic materials.

The rate of cellulose degradation is also strongly affected by end-product inhibition caused by cellobiose and glucose, cellobiose being a stronger inhibitor in the hydrolysis reaction as observed by Lee *et al* (3). To minimise inhibition by cellobiose and to maximise glucose formation, it is necessary to use cellobiase along with the cellulase application. Increasing the amount of cellobiase in the cellulase mixture may lower or prevent the accumulation of inhibitory cellobiose, thereby, increasing the yield of glucose as well as enhancing the rate of reaction.

With this background in mind, the present work has been undertaken to study the effects of swelling and of using mixed enzymes for improved yields of fermentable sugars by saccharification.

2. MATERIALS AND METHODS

Oil palm press fibres used in this study were collected from Malpom Industry, Penang. After washing and air-drying, it was ground to an average particle size of less than 1 mm using 'Starch Plant Shredder'. Enzymes used were cellulase and cellobiase both from Novo with activities of 1500 and 250 IU/ml respectively.

2.1 Soaking

Oil palm press fibres were first soaked in NaOH solutions with concentrations varying from 0.5 to 2.0 M (w/v) and for the periods ranging from 2 to 48 h. The substrate was then autoclaved for 15 minutes at 121°C followed by filtering and thorough washing until the alkali was completely removed. The product was then dried and the loss in weight in each case was determined. Swelling was in direct proportion to the loss in weight with alkali.

2.2 Boiling

For this, PPF was put under boiling (100°C) at different concentrations of NaOH (0.5 to 2.0 M) for 2 h and after the period, the procedures for washing, drying and taking weight loss as in soaking were followed. In both soaking and boiling, the ratio between the substrate (treated PPF) and the alkali was 1:20 (w/v).

2.3 Hydrolysis

The solid concentration maintained for saccharification by enzyme was 1% (w/v) using 0.5M acetate buffer at pH 4.8. Experiments were first conducted with cellulase only (using 18 IU/ml) in the shaking incubator (LH Fermentation, Model F2000, Germany) under controlled conditions of temperature and agitation at 48°C and 150 rpm respectively as followed by Dhawan and Gupta (4). Samples were collected periodically, centrifuged and the supernatant was analysed for reducing sugars following DNS method after Miller (5). Experiments with mixed enzymes of cellulase and cellobiase with activities of 18 and 5.4 IU/ml respectively were also performed under similar optimum conditions and RS was measured at regular intervals.

3. RESULTS AND DISCUSSION

3.1 Effect of Swelling on Hydrolysis

Results on saccharification of PPF after delignification by the processes designated in this study as 'boiling' and 'swelling' are presented in Figure 1. The figure shows that the rate of hydrolysis increases with increase in swelling upto a certain limit (when PPF was treated with 1M NaOH for 2 h with 53% weight loss, Table 1), beyond which the rate decreases. The conversion attained after 48 h using swelling process is about 58% and that by boiling method is 47% during the same period of hydrolysis by cellulase. Thus, it is found that swelling has caused about 10% increase in conversion.

The initial rate of hydrolysis is proportional to the swelling, thereby, forming a straight-line relationship. Beyond the above limit (after 53% weight loss), though the percentage of cellulose increases, the rate decreases. This may be due to the fact that at higher alkali concentrations, oxidation may occur changing the -CH_2OH group of the cellulose to -COOH forming polyglucuronic acid. By further decarboxylation, this may form xylan, thus, changing the nature of the cellulose molecule.

Summarised results on the extent of hydrolysis (% conversion of treated PPF into reducing sugars) after 24 h are presented in Table 2. The conversion of PPF whether boiled or swelled is found to be in the range of 7- 8 % after 24 h and the same reached its maximum (48.3 and 56% for boiling and swelling respectively) after 96 h. When treated with enzyme, conversion in both cases after 24 h is observed to be around 11%, whereas, for both, maximum conversions (67 and 83%) have reached after 96 h. No more conversion could be attained after this. In short, it is observed that swelling gives conversion as high as 83%.

However, removal of lignin from palm fibres is found to be very effective for subsequent saccharification. These observations correlated with weight loss (in terms, 'swelling') of solid waste after each alkali treatment are shown in Table 1.

The enhancement in the hydrolysis rate is possibly due to the removal of lignin and structural swelling of the substrate. It is known that cellulose is suseptible to weak alkaline solutions at temperatures higher than 100°C and that some depolymerization of the carbohydrates takes place during the delignification as shown by Casey (6). This depolymerization of cellulose may also be responsible for this increase in hydrolysis rate.

The increased yield on swelling may also be due to the fact that cellulose assumes a better character with larger voids which increases the mobility or diffusibility of cellulose into the swollen substrate causing higher yield as observed by Han (7). This can be seen in Figures 2-5, which are the pictures where the substrate was observed under scanning electron microscope (SEM) before and after pretreatments. From Figure 1, it is seen that there is an insignificant structural swelling when PPF is untreated. When treated with 0.5, 1.0 and 2.0 M NaOH for swelling, the structural views under SEM are shown in Figures 3, 4 and 5 respectively. It is observed from the figures that maximum structural swelling of PPF occurs when 1.0 M NaOH is used (Figure 4) which also gives the highest conversion.

3.2 Effect of Supplemented Cellobiase on the Hydrolysis of Oil Palm Press Fibres

Cellobiase which hydrolyses cellobiose to glucose is known to be inhibitory to both product and substrate. Although cellobiase was not active on the cellulose itself, its addition has greatly increased the rate and extent of hydrolysis by ensuring the efficient hydrolysis of cellobiose and reducing the influence of end-product inhibition.

In this study, cellobiase was supplemented with cellulases in the process of saccharification. Results on the effect of cellobiase addition are shown in Figure 6. It is seen from the results that the addition has resulted in 60% conversion at 24 h and as high as about 90% within 48 h, which are higher when compared with those using only cellulase (44 and 58% conversion at 24 and 48 h respectively). But, under optimised swelling conditions (soaking for 4 h in 1.5 M NaOH), the highest conversion obtained was 76% with cellulase. Thus, the addition of cellobiase for saccharification is found to further enhance the production of total sugars (to about 14% increase in conversion).

TABLE 1. Effect of the pretreatments on the weight loss of the substrate.

Pretreatments	Percent weight loss (based on total weight loss)
Soaking (2 h) in	
0.5 M NaOH	42
1.0 M NaOH	53
1.5 M NaOH	55
2.0M NaOH	64
Boiling (2 h) in	
0.5 M NaOH	26
1.0 M NaOH	30
1.5 M NaOH	35
2.0 M NaOH	50

The significant loss in weight may be due to the degradation of hemicellulose and the solubilisation of part of the lignin.

TABLE 2. Effect of boiling and swelling on the rate of hydrolysis.

Pretreatments (Boiling and swelling in 1 M NaOH for 2 h)	Extent of hydrolysis after 24 h (Conversion, %)	Extent of hydrolysis after 96 h (Conversion, %)
Boiling + C	7.1	48.3
Swelling + C	8.3	56.4
Boiling + C + CB	11.3	67.2
Swelling + C + CB	11.4	83.1

C: cellulase
CB : Cellobiase

The incomplete conversion by cellulase may be explained by the fact that the accumulation of cellobiose has inhibited the production of glucose. After reaching a maximum conversion between 24 and 48 h, the cellobiose concentration slowly decreases, resulting in a corresponding increase in the glucose concentration.

ACKNOWLEDGEMENTS

The authors are grateful to 'Malayan Sugar Manufacturing (MSM) Co. Bhd.' for their financial support in this work.

REFERENCES

1. Fan, L. T, Lee, Y. H and Gharpuray, M. M, 1982. Adv. Biochem Eng., 23, 157
2. Ladisch, M. R, Ladisch, C. M and Tsao, G. T, 1978. Science, 201, 743
3. Lee, Y. H, Gharpuray, M. M and Fan, L. T, 1982. Biotechnol. Bioeng.Symp., 12, 121
4. Dhawan, S and Gupta, T. K, 1977. J. Gen. Appl. Microbiol., 23,15
5. Miller, G. L, 1959. Anal. Chem., 31, 426
6. Casey, J. P, 1960. Pulp and Paper, Vol. 1, Interscience, New York
7. Han, Y. W, 1978. Adv. Appl. Microbiol., 23, 119

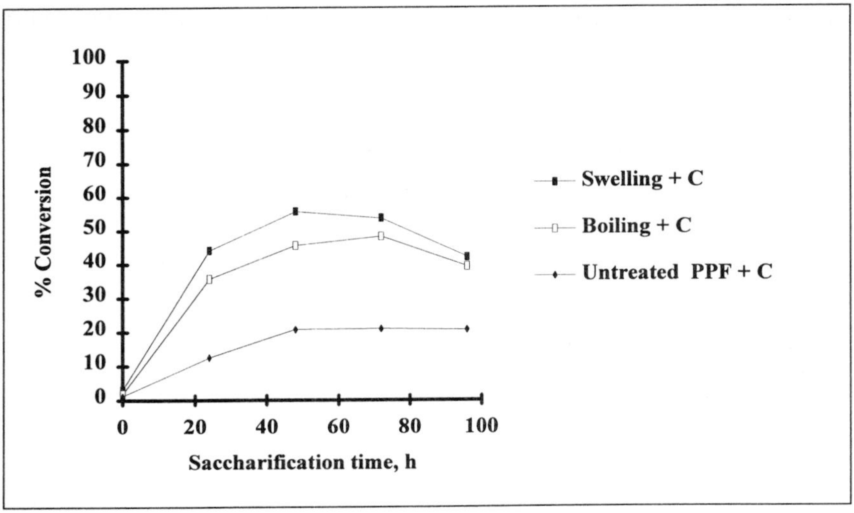

Fig. 1. Effect of different treatments on the hydrolysis of treated PPF under optimum conditions. [pH 4.8, temperature of 48°C, boiling in 1M NaOH at 100°C for 2 h, swelling in 1M NaOH for 2 h]. C: cellulase, 18 IU/ml substrate volume.

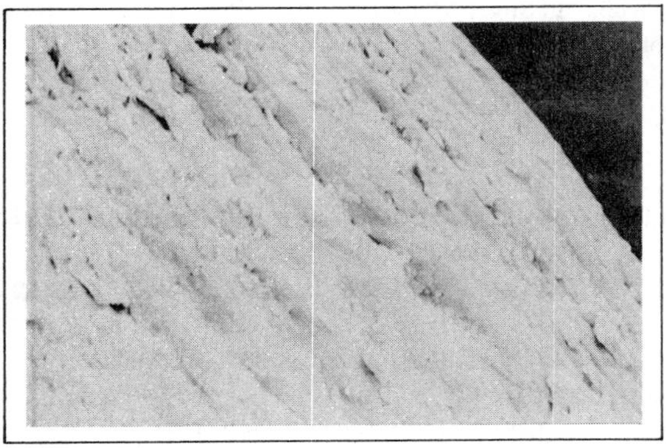

Fig. 2. Microstructural view of untreated PPF under scanning electron microscope (SEM). Mag. 2000X

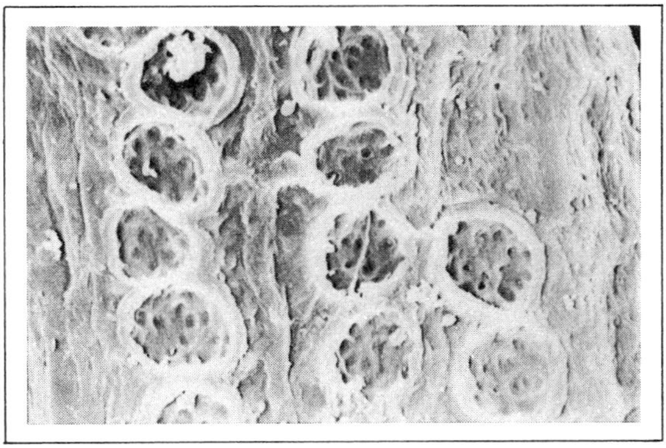

Fig. 3. SEM of treated PPF with 0.5 M NaOH. (Mag. 2000X)

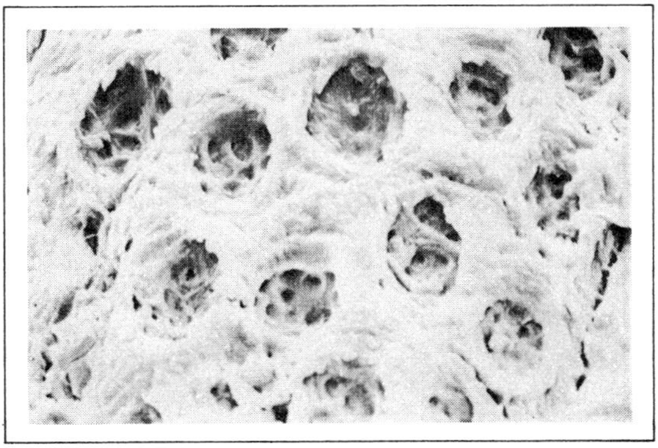

Fig. 4. SEM of treated PPF with 1.0 M NaOH. (Mag. 2000X)

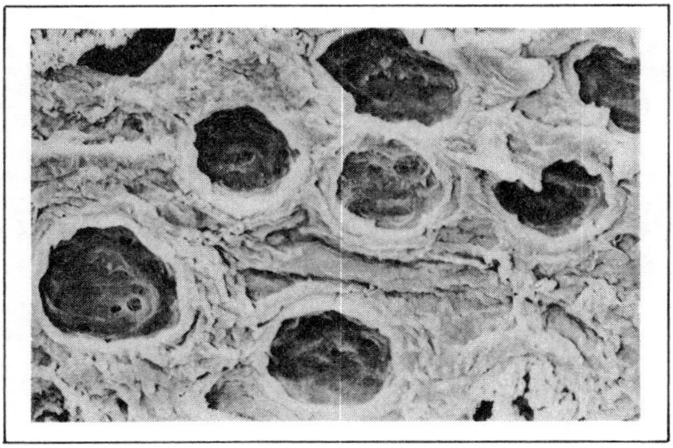

Fig. 5. SEM of treated PPF with 2.0 M NaOH. (Mag. 2000X)

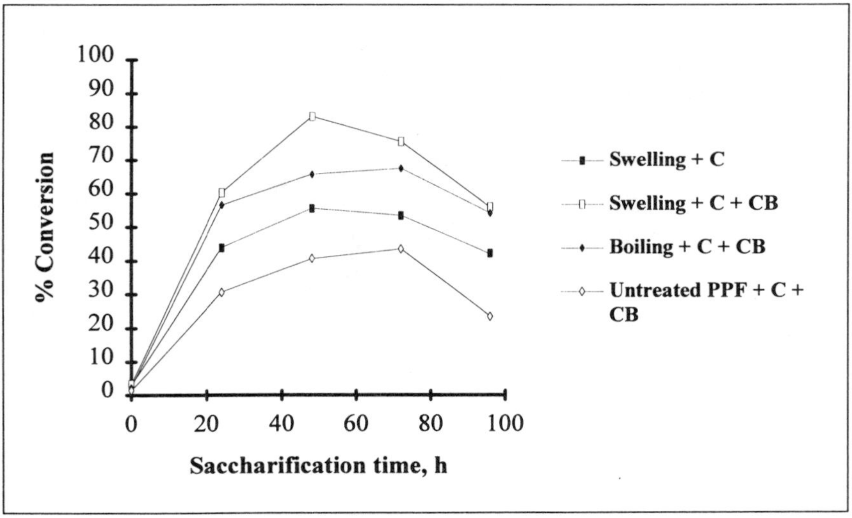

Fig. 6. Effect of supplemented cellobiase on the hydrolysis of treated PPF under optimum conditions. [pH 4.8, temperature of 48°C, boiling in 1M NaOH at 100°C for 2 h, swelling in 1M NaOH for 2 h]. C:cellulase, 18 IU/ml substrate volume. CB: cellobiase, 5.4 IU/ml substrate volume.

SEPARATION TECHNIQUES IN INDUSTRIAL BIOPROCESSING

Yusuf Chisti and Murray Moo-Young
Department of Chemical Engineering, University of Waterloo, Waterloo, Ontario, Canada N2L 3G1

An overview of considerations for developing industrial bioseparation schemes is presented with emphasis on integration of the various separation operations into an overall downstream purification process. Purification of extracellular and intracellular products is discussed, as well as recovery of recombinant proteins.

1. INTRODUCTION

Raw products of microbial fermentations, animal or plant cell cultures and enzyme bioreactors invariably undergo one or more separation and purification steps to recover the product in the desired form, concentration and purity. Processing beyond the bioreaction step is termed downstream processing which consists of predominantly physical separation operations. Some commonly used operations are listed in Table 1.

The steps of a properly engineered downstream process are integrated with each other and with the bioreaction stage to yield an optimal scheme for recovery of the product. This discussion is limited to factors which must be considered in developing any economically viable product purification and concentration scheme based on a small selection of the many downstream processing operations that are available. Individual operations have been described elsewhere (1-4).

2. DESIGN OF A BIOSEPARATION SCHEME

2.1 General Considerations

Design of a bioseparation scheme and the engineering of the process equipment require careful consideration of the physical and chemical

Table 1. Bioseparation operations.

Solid-liquid separations (1-4)	*Chromatographic methods* (1-4,5)	*Miscellaneous*
Sedimentation	Gel permeation	Adsorption
Filtration	Ion exchange	Electrophoresis
Centrifugation	Hydrophobic interaction	Cell disruption (7,8)
Flotation	Affinity	Crystallization
		Flocculation
		Extraction
		Liquid-liquid extraction (2,9)
Thermal operations	*Membrane separations* (1-4)	
		Precipitation (2)
Distillation	Microfiltration	
Evaporation	Ultrafiltration	
Drying (2)	Pervaporation (6)	
	Reverse osmosis	

properties of the material being handled and delimitation of the maximum processing stresses (temperature, pH, shear forces, contamination, pressure, exposure to denaturing chemicals) that the material may safely tolerate (2). Typically, a separation process must operate within the physiological ranges of pH and temperature (pH ~ 7.0; temperature ≤ 37° C), the specific conditions being very process dependent. For example, enzymes such as lysozyme, ribonuclease and acid proteases are quite stable at low pH values. Sometimes during processing, exposure of material to relatively severe environmental conditions is unavoidable. In such cases, the duration of exposure is minimised and other precautions (*e.g.*, low temperature; addition of chemicals to reduce oxidation, *etc.*) are taken to reduce the impact of exposure. Some biologically active molecules, particularly proteins, may be sensitive to excessive agitation; however, enzymes, with the exception of multienzyme complexes and membrane-associated enzymes, are not damaged by shear in the absence of gas-liquid interfaces (2,10).

The first few processing operations in a purification train are aimed at volume reduction to minimise processing costs by reducing the size of the downstream machinery. Removal of suspended material and substances which might interfere with further downstream operations are additional requirements of some of the early separation steps. Further, because viscous broths are difficult to handle, viscosity reduction should be achieved as early as possible to simplify pumping, mixing, filtration, *etc*. Removal of suspended solids, digestion of carbohydrates, or removal of nucleic acids are some of the operation that may be needed to improve broth handling.

Typically, solid-liquid separation would be among the first processing steps for extracellular as well as intracellular products. For the latter, solid-liquid separations are usually a means of concentration of the biomass, or removal of the suspending culture fluid prior to disruption or other downstream treatment. Cell or other solid product washing operations often employ solid-liquid separation steps. The commonly used methods of solid-liquid separation are filtration and centrifugation. These methods can be implemented in a variety of forms which are best suited to particular applications. Thus, as detailed in Table 2, many different designs of centrifuges are available (1,2). Similarly, filtration may be performed in conventional filter presses, rotary drum pressure or vacuum filters with or without filter aids and using different means of solids discharge. Alternatively, membrane filtration may be used either in dead end or cross flow modes; the latter may be implemented in flat plate, hollow fibre or spiral wound membrane cartridges (2). While the variety of available options helps to ensure that specific need are met, careful consideration of the problem at hand is required for selection of the optimal processing method. Alternatives should be considered whenever possible. For example, rotary drum filters with string discharge usually perform well in separating mycelial solids from penicillin broths, but this discharge mechanism, without filter aids, causes problems with broths of *Streptomycetes*. Similarly, because of the concentration and the morphology of the solids, the disc stack centrifuge is not suitable for fungal fermentation broths, but properly selected scroll discharge machines are effective. Gravity sedimentation may be employed as a volume reduction step prior to removal of solids by other means, but sedimentation by itself is not common for biomass removal in processing of high value products.

When more than one processing options are technically feasible, evaluations of the economics of use in terms of capital expenditure on equipment and its operating costs (processing time, yields, labour, cleaning, maintenance, analytical support) is necessary for optimal process selection. Economic evaluations should be performed over the expected lifetime of the equipment. For example, for separation of solids from fermentation broth, centrifugation and microfiltration may be two competing alternatives (2). In still other applications, for example when very fragile cells are to be separated from suspending liquid, centrifugation may not at all be an option.

Some other concentration steps, applicable to products in solution, are precipitation (2,11), adsorption, chromatography (5), evaporation, pervaporation (6) and ultrafiltration (2). Some of these operations are equally capable purification steps (*e.g.*, chromatographic separations). Certain steps (*e.g.*, some chromatographic separations; membrane separations) may require a relatively clean process stream, free of debris, lipids or micelles which may cause fouling of the equipment. Such steps are often used downstream of steps which can handle cruder material.

Table 2. Types and applications of centrifuges.

Tubular bowl. Tubular bowl machines are capable of high g-forces; solids accumulate in the bowl and must be removed manually at the end of operation. Bowl capacity limits solids-holding capability. To ensure sufficient interval between bowl cleaning, the solids concentration in the feed should usually be ≤ 1% volume/volume; higher concentrations can be processed with smaller batches. Good dewatering of solids is obtained.

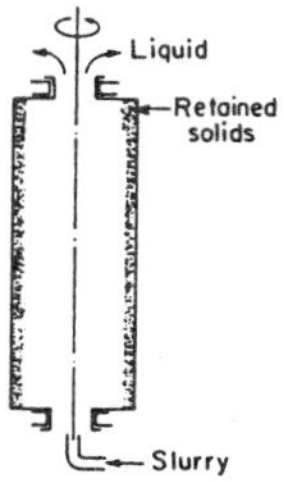

Multichamber bowl. Similar to tubular bowl machines. Division of bowl into multiple chambers increases solids holding capacity. Solids must be discharged manually; hence, economic operation is feasible only with feeds with low concentration of solids. Good for polishing of otherwise clarified liquors. Capable of high g-forces. Gradation of g-forces from inner to outer chamber. Smallest particles sediment in the outermost chamber. Good dewatering of solids.

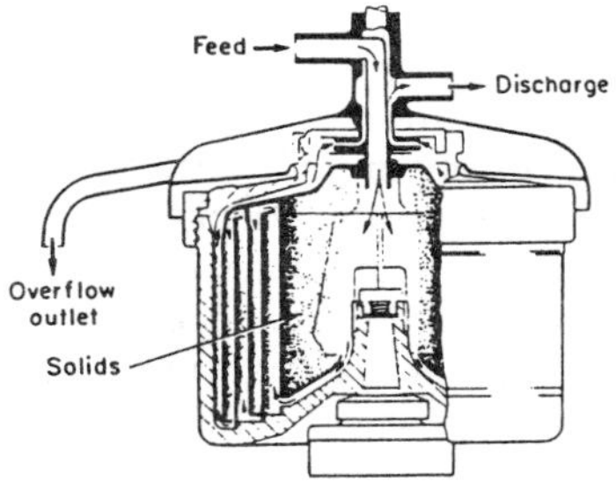

Disc-stack. Lower g-forces than tubular bowl machines. Solids may be retained, or discharged intermittently or continuously by various mechanisms. Solids must flow. Poor dewatering. Not suited for mycelial solids; good for slurries of yeasts and certain bacteria. Depending on the mechanism of solids discharge, may handle feeds with up to 30% (v/v) solids.

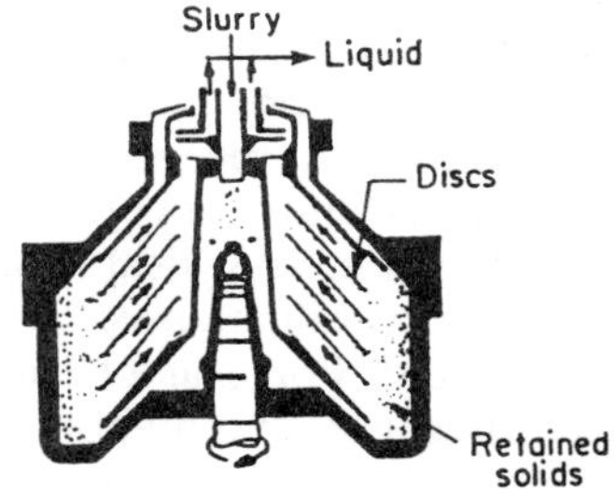

Scroll discharge. Scroll discharge decanter centrifuges are suitable for slurries with high concentration of relatively large, dense solids. Feed solids concentrations of 5-80% (v/v) can be handled. Solids are discharged continuously. The g-forces are low.

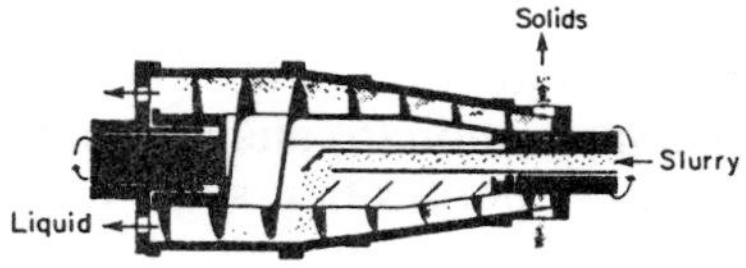

Sometimes the characteristics of fermentation broth or process liquor may be modified by pretreatment to enable processing by a certain method. Treatments such as pH adjustments and addition of polyelectrolytes or other flocculants can be used to achieve major changes in processing characteristics.

As far as possible, the requisite purification and concentration should be achieved with the fewest processing steps; generally, no more than 6-7 steps are used, a situation quite different from that in chemistry and biochemistry laboratories, where the number of individual steps is often not a major consideration, purity of the product is usually more important than overall yield or costs. A train of only five steps, each with 90% step yield, would reduce the overall recovery to less than 60% (12). To minimize reduction of the overall yield, high resolution separations such as chromatography should be utilized as early as possible in the purification scheme in keeping with the processing constraints that these steps require (*e.g.*, clean process streams free of debris, particulates, lipids, *etc.*). Commonly used sequences of protein purification methods have been discussed (11).

Separation schemes incorporating unit operations which utilize different physical-chemical interactions as the bases of separation are likely to achieve the greatest performance for a given number of steps. Combining two separation stages based on the same separation principle may not be an effective approach. As an example, when two chromatographic steps in series are selected, gel filtration which separates based on molecular size and ion exchange chromatography which separates based on difference in charge on the molecules, may be a suitable combination.

Except for the final few finishing operations, downstream processing is usually conducted under non-sterile, but bioburden controlled, conditions; however, prevention of unwanted contamination and cleaning and sanitization considerations require that the processing machinery be designed to the same high standards as have been described for sterile bioreactors (13,14).

2.2 Intracellular Products

In general, a biological product is either secreted into the extracellular environment, or it is retained intracellularly. In comparison with the total amount of biochemicals produced by the cell, very little material is usually secreted to the outside; however, this selective secretion is itself a purification step which simplifies the task of the biochemical engineer. Extracellular products, being in a less complex mixture, are relatively easy to recover. On the other hand, because a greater quantity and variety of biochemicals are retained within cells, intracellular substances are bound to eventually become a major source of bioproducts (7). Among some of the newer intracellular products are recombinant proteins produced as dense inclusion bodies in bacteria and yeasts. Recovery of intracellular products is more expensive as it

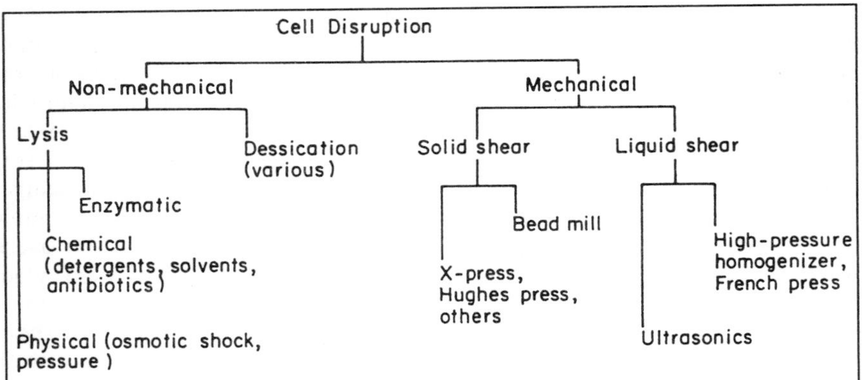

Figure 1. Cell disruption alternatives (7).

requires such additional processing as cell disruption (7,8), lysis (7) or permeabilization (15). Intracellular polymers such as polyhydroxybutyrate (PHB) may be recovered either by cell disruption (8) or solvent extraction. In principle, selective release of the desired intracellular products is possible, but in practice it is neither easily achieved nor sufficiently selective. Hence, the desired product must be purified from a relatively complex mixture, complicating processing and adding to the cost. Nevertheless, an increasing number of intracellular products are in production. Economics of production may be improved by recovering several products (intracellular and extracellular) from the same fermentation batch (10).

As for other separations, many options exist for disruption of cells (Figure 1). Of these, high pressure homogenization is apparently the most suitable for bacterial broths, whereas bead mills are more widely used for fungal cultures (2,7). Disruption of bacterial cells releases large amounts of nucleic acids which increase the viscosity of the broth, often producing viscoelastic behaviour. To ease further purification, the nucleic acids are usually removed by precipitation (*e.g.*, with manganous sulfate, streptomycin or polyethyleneimine) (2); alternatively, viscosity may be reduced by enzymatic digestion of nucleic acids or high-shear processing in high pressure homogenizers.

2.3 Impact of Product Specifications

The specifications on product purity and concentration should be carefully considered in developing a purification protocol. Concentration or purification to levels beyond those dictated by needs is wasteful. The acceptable level of contamination in a particular bioproduct depends on the dosage, the frequency of use and the method of application (*e.g.*, food, drug, oral, parenteral), as well as on the nature and toxicity (or perceived risk) associated with the contaminants. Products such as vaccines, which are used only a few times in a lifetime, may be acceptable with relatively high levels of other than the desired biomolecule. In some cases, contaminating protein levels of about 100 ppm may be acceptable. The purification process must also reduce the nucleic acids from recombinant and other cells to low levels. An acceptable nucleic acid concentration of $\leq 10 \times 10^{-15}$ kg per dose has been suggested. In addition, parenteral products, other than certain vaccines, must be free of microorganisms and viruses.

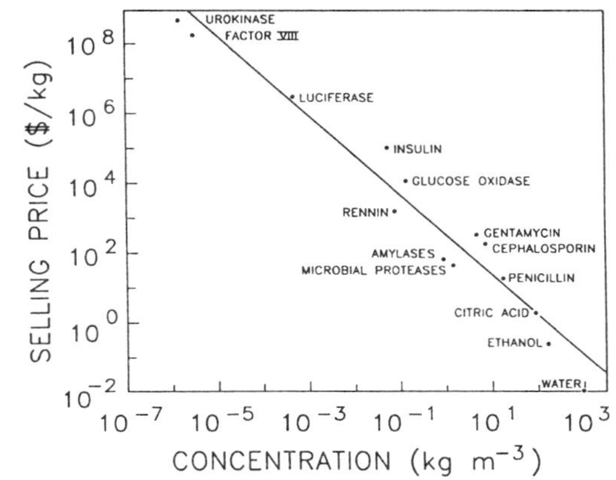

Figure 2. Cost of some bioproducts vs. concentration in the raw fermentation broth. Based on Dwyer (16).

Products derived from bacteria such as *Escherichia coli* will invariably be contaminated with bacterial cell wall endotoxins which can cause adverse reactions (headaches, vomiting, diarrhoea, fevers, *etc.*) in patients unless reduced to very low levels (*e.g.*, less than 5×10^{-13} kg per kg body weight). *In vitro* diagnostic proteins (enzymes, monoclonal antibodies) may tolerate greater levels of contaminants so long as the contaminants do not interfere with the analytical performance of the product. With certain diagnostic proteins, such as the blood typing monoclonal antibodies, cross-contamination causing misdiagnosis is an extreme concern because of possibly fatal consequences of mis-typing. Such concerns influence the design and operation of the downstream process, particularly for multi-product plants.

For many biological products, particularly pharmaceuticals, seemingly minor alterations in downstream processing can have important implications on the performance of the product. For example, penicillins may be recovered by liquid-liquid extraction of either the whole fermentation broth or solids-free broth. The latter scheme requires an additional solid-liquid separation step than the whole broth process. However, the whole broth extracted product has been known to cause more frequent cases of allergenic reactions in comparison with the other processing alternative. In fact, some pharmaceutical companies now demand of contract suppliers that, in addition to meeting product specifications in terms of measurable contamination, the product they supply must conform to a certain production method, in this case extraction after removal of fungal solids. When raw penicillin is for bulk conversion to semi-synthetic penicillins, whole broth extraction may be acceptable in view of the security afforded by the additional steps involved in making and purifying 6-aminopenicillanic acid from raw penicillin.

2.4 Impact of the Bioreaction Step

Downstream processing typically represents 60-80% of the cost of production of fermentation products. Thus, superficially it may appear that process improvement should focus on downstream. This is not so. Even small improvements in the yield or purity of the product in the bioreaction step can have a significant effect on downstream recovery costs. This is clearly shown in Figure 2, where the cost of production (reflected in selling price) of several biochemicals is plotted as a function of the product concentration in the fermentation broth or the starting material. The potential for yield improvement at the bioreaction stage is usually high. Major yield enhancements have been fairly commonly achieved by strain selection, medium development, optimization of feeding strategies and environmental controls. For some processes, alternative microorganisms may be a viable option. Preference should be given to faster growing, easy to process organisms. Process improvement or intensification should emphasize a global approach. Schemes which combine the bioreaction stage and parts of downstream processing, for example extractive fermentations, or fermentation-distillation, can potentially improve overall process yield and economics; however, so far, these schemes have received little commercial attention.

3. RECOMBINANT PROTEINS

Many of the newer recombinant biotechnology products are proteins. While the general features of a bioseparation scheme for these products are the same as for other proteins, there are some unique constraints. Genetically modified microorganisms and cells of higher life forms are often more fragile than the corresponding wild strains (17,18). This has implications for the design of cell-liquid separation stages. Also, recombinant proteins formed in bacteria and yeasts frequently precipitate inside the cell as dense, insoluble, denatured, inclusion bodies. In this form proteins which may otherwise be toxic to the cell may be overproduced and remain protected against proteolytic activity within the cell.

Most bacteria and fungi used in producing recombinant proteins also produce a variety of proteases. The proteolytic activity may degrade some of

Table 3. Some proteins produced as inclusion bodies.

Bovine growth hormone	Human macrophage-colony stimulating factor	Prochymosin
Bovine pancreatic ribonuclease	Human serum albumin	Pro-urokinase
Epidermal growth factor	Immunoglobulins	Tissue-type plasminogin activator
Human interleukin-2	Lysozyme	
Human interleukin-4	Porcine phospholipase	

the desired protein within the cell and during recovery. Soluble, non-inclusion body, proteins being particularly susceptible to degradation. Degradation by acid proteases with a pH optimum of 2-4 may be minimized by processing at higher pH and low temperatures. Neutral proteases are not particularly thermostable and may be inactivated by heating to 60°-70°C for 10-minutes. Many proteases are metaloproteins and require a divalent metal ion for proteolytic activity; chelating agents such as ethylenediaminetetraacetic acid (EDTA) or citric acid may be used to inactivate such proteases by binding the metal ions. Alkaline proteases of *Bacilli*, such as subtilisin, contain serine at the active site and are not affected by EDTA, but are inhibited by diisopropylfluorophosphate. The short lived reagent phenylmethylsulfonyl fluoride protects against serine proteases. Antioxidants such as Vitamin E and ascorbic acid protect against oxidation.

Proteins tend to be more stable in concentrated solutions. Addition of polyethylene glycols and other proteins such as albumins may have an stabilizing effect. Glycerol, sucrose, glucose, lactose and sorbitol are often used as stabilizers in concentrations of 1-30%. Enzyme substrates usually have an stabilizing effect as do high concentration of salts such as ammonium sulfate and potassium phosphate. Metaloproteins may be stabilized by addition of metal salts. Divalent metal ions such as Ca^{2+}, Cd^{2+}, Mn^{2+}, Zn^{2+} stabilize various enzymes.

3.1 Inclusion Body Proteins

When possible, production of recombinant proteins as inclusion bodies has important advantages. Some proteins which form inclusion bodies are listed in Table 3. Inclusion bodies are easy to isolate, highly concentrated forms of the desired recombinant protein. Typically, inclusion bodies are spheroidal particles, 0.2-2.0 × 10^{-6} m in diameter and 1100 - 1300 kg•m^{-3} density. The sequence of steps in recovery of inclusion body proteins is cell disruption, centrifugal separation of the inclusion body, washing, solubilization of the protein and renaturation. Cell disruption by homogenization is the preferred technique in large scale processing. Disruption by high pressure homogenization has been detailed by Chisti and Moo-Young (7). Inclusion bodies are not affected by homogenization. Cell homogenates are centrifuged to sediment the dense inclusion body fraction. Centrifugation at 1000-12,000-g for 3-5 minutes is sufficient. Sedimentation of cell debris can be minimized by increasing the density and viscosity of the homogenate by addition of sucrose. The inclusion body fraction is washed with buffers containing 1M sucrose, 1-5% Triton-100 surfactant (19) and, in some cases, low concentrations of proteolytic enzymes and denaturants. The wash steps removes soluble contaminants, membrane proteins, lipids and nucleic acids. At this stage the remaining solids fraction is >90% recombinant protein. The protein solids are solubilized in highly denaturing chaotropic media. Typically, 6-8 M guanidine hydrochloride or 8M urea are used for solubilization at pH 8-9, 25-37°C for 1-2 hours (19). Reducing agents are added to the solubilization media to break any inter- and intra-molecular disulfide bonds to fully solubilize the protein. Some reducing agents are 2-mercaptoethanol, dithiothreitol, dithioerythritol, glutathione, 3-mercaptopropionate. Some typical concentrations are 0.1M 2-mercaptoethanol, or 10 mM dithiothreitol (19). The latter has a shorter half-life than 2-mercaptoethanol, but does not have the odour of 2-mercaptoethanol. Stability of thiol compounds in solution is dependent on pH, temperature and the presence of such metal ions as Cu^{2+}, which lower stability, and of stability enhancers such as EDTA. Good yields of some proteins can be obtained by solubilization without the reducing reagents, but for others reducing agents are essential. Of the denaturants, guanidine hydrochloride is preferable to urea which may contain cyanate causing carbamylation of the free amino groups on the protein, particularly during long incubation periods in alkaline environments. Note though, that for some proteins, one denaturant may produce significantly higher overall yield than if solubilization with the other is used. Performance has to be empirically evaluated.

For refolding of solubilized protein into active entities, concentration of the denaturant and the reducing agent are reduced by dilution with a refolding buffer. Denaturants can be completely removed by ultrafiltration with addition of renaturing buffer, dialysis or gel filtration. Renaturation from concentrated protein solutions produces lower yields of the active protein because of intermolecular aggregation in these solutions. Thus, renaturation is done at low protein concentrations, typically $1\text{-}20 \times 10^{-3}$ kg•m^{-3} protein (19). Yield of the active protein is enhanced by refolding in the presence of small, non-denaturing, amounts (1-2 M) of urea or guanidine hydrochloride (19). Presence of high molecular weight polymers such as polyethyleneglycol may also improve yield.

During refolding, formation of the disulfide bonds is achieved by one of three ways. The air oxidation method uses dissolved oxygen for oxidation of the cystine residues. The refolding buffer containing solubilized protein may be aerated or exposed to atmosphere. Oxidation is accelerated by Cu^{2+} ions at approximately 10^{-6} M. Typical reaction conditions are pH 8-9, 4-37° C for up to 24-hours (19). Traces of 2-mercaptoethanol may enhance yield. Air oxidation is difficult to control.

The glutathione reoxidation method typically uses a 10:1 mixture of reduced and oxidized forms of glutathione at a concentration of 10^{-3} M reduced glutathione (19). Air oxidation is suppressed by using deaerated buffers held under a nitrogen atmosphere. The ratio of the reduced and oxidized forms of glutathione, the ratio of the glutathione and the cystine residues on the protein, the reoxidation temperature (4-37° C) and time (1-150 hours) provide flexibility to this method (19). Low molecular weight thiols other than glutathione may also be used. The third method of disulfide bond formation, the mixed disulfide interchange technique, has been detailed by Fischer (19).

The inclusion body production stage should be optimized to rapidly form relatively pure, large and dense inclusion bodies which are easy to recover and solubilize. Production of proteolytic activity should be suppressed as far as possible. Purification and concentration are greatly simplified because of the already high starting protein concentration and purity in the inclusion bodies which are easy to separate from the bulk of the soluble proteins by centrifugation. The recovery of active protein from inclusion bodies is variable, but can approach 100%. In general, smaller polypeptides are easier to refold into active forms.

4. CONCLUSIONS

Recovery of bioproducts at acceptable cost requires attention to selection of suitable separation and purification methods integrated with each other and with the bioreaction stage. Fewest possible individual processing steps, consistent with the product quality specifications, should be used.

5. REFERENCES

1. Moo-Young, M., editor, 1985, *Comprehensive Biotechnology*, vol. **2**, Pergamon Press, Oxford.
2. Chisti, Y. and Moo-Young, M., 1991, In: Moses, V. and Cape, R. E., editors, *Biotechnology: The Science and the Business*, Harwood Academic Publishers, New York, pp. 167-209. Fermentation technology, bioprocessing, scale-up and manufacture.

3. Belter, P. A., Cussler, E. L. and Hu, W.-S., 1988, *Bioseparations: Downstream Processing for Biotechnology*, John Wiley, New York.
4. Wheelright, S. M., 1991, *Protein Purification: Design and Scale up of Downstream Processing*, Hanser Publishers, New York.
5. Chisti, Y. and Moo-Young, M., 1990, *Biotechnol. Adv.*, **8**, 699-708. Large scale protein separations: Engineering aspects of chromatography.
6. Fleming, H. L., 1992, *Chem. Eng. Prog.*, **88**(*7*), 46-52. Consider membrane pervaporation.
7. Chisti, Y. and Moo-Young, M., 1986, *Enz. Microbial Technol.*, **8**, 194-204. Disruption of microbial cells for intracellular products.
8. Harrison, S. T. L., 1991, *Biotechnol. Adv.*, **9**, 217-240. Bacterial cell disruption: A key unit operation in the recovery of intracellular products.
9. Abbott, N. L. and Hatton, T. A., 1988, *Chem. Eng. Prog.*, **84**(*8*), 31-41. Liquid-liquid extraction for protein separations.
10. Dunnill, P., 1983, *Process Biochem.*, **18**(*5*), 9-13. Trends in downstream processing of proteins and enzymes.
11. Bonnerjea, J., Oh, S., Hoare, M. and Dunnill, P., 1986, *Biotechnology*, **4**, 954-958. Protein purification: The right step at the right time.
12. Fish, N. M. and Lilly, M. D., 1984, *Biotechnology*, **2**, 623-627. The interactions between fermentation and protein recovery.
13. Chisti, Y., 1992, *Chem. Eng. Prog.*, **88**(*1*), 55-58. Build better industrial bioreactors.
14. Chisti, Y., 1992, *Chem. Eng. Prog.*, **88**(*9*), 80-85. Assure bioreactor sterility.
15. Dörnenburg, H. and Knorr, D., 1992, *Process Biochem.*, **27**, 161-166. Release of intracellularly stored anthraquinones by enzymatic permeabilization of viable plant cells.
16. Dwyer, J. L., 1984, *Biotechnology*, **2**, 957-964. Scaling up bioproduct separation with high performance liquid chromatography.
17. Dunnill, P., 1987, *Chem. Eng. Res. Des.*, **65**, 211-217. Biochemical engineering and biotechnology.
18. Moo-Young, M. and Chisti, Y., 1988, *Biotechnology*, **6**, 1291-1296. Considerations for designing bioreactors for shear-sensitive culture.
19. Fischer, B. E., 1994, *Biotechnol. Adv.*, **12**, 89-101. Renaturation of recombinant proteins produced as inclusion bodies.

PRODUCTION OF CARBOHYDRATE-BASED FUNCTIONAL FOODS USING ENZYME AND FERMENTATION TECHNOLOGIES

Martin J. Playne, CSIRO Division of Food Science and Technology, Dairy Research Laboratory, Melbourne, Australia

Oligosaccharides act as functional foods. Food-grade oligosaccharides are produced either by controlled hydrolysis of plant polysaccharides or by synthesis from sugars using hydrolase enzymes acting by transglycosylation. Methods of production are described for fructo-, galacto-, xylo-, inulo- and iso-malto oligosaccharides, and commercial data given. Current Australian research on oligosaccharides is reviewed; the patent literature of 260 patents in the last five years summarised; and current and proposed uses and applications for oligosaccharides are discussed.

1. INTRODUCTION

Functional foods are defined as foods which are derived from naturally-occurring substances which can and should be consumed as part of the daily diet, and which have a particular function when ingested by serving to regulate a particular body process. Officially, they are termed 'Foods for Specified Health Use' (FOSHU) by the Japan Ministry of Health and Welfare (Van den Broek(1)). Functional foods are also known as "healthy foods". They are not to be confused with the term "functionality" which is a term used by food processors to indicate physico-chemical properties of a food ingredient leading to certain processing properties when the ingredient is added to a food. The terms "pharma foods", "nutriceuticals", "therapeutic foods" refer to particular types of functional foods which confer both nutritional and pharmaceutical properties.

Oligosaccharides are carbohydrates of three to ten monomer sugars in size. They include gluco-, iso-malto-, galacto-, xylo-, and fructo-oligosaccharides, and can act as an essential component of functional foods. Disaccharides, such as lactulose and galactobiose, also exhibit some similar functional properties, and are sometimes considered to be oligosaccharides.

Oligosaccharides are found as major components of many natural products (e.g. plant cells, milk) in either free or combined form. The importance of their presence in most plant and animal cells attached to proteins has been recognised recently. Their ability to act as specific recognition molecules in cell biology has led to the emergence of the science of glycobiology (Rademacher *et al*(2)). This paper is not about glycobiology. It is concerned with the role of mixtures of oligosaccharides as functional foods. Oligosaccharides can act in a number of ways in foods (see Table 1), but a most important role is their ability to pass indigested through the stomach and small intestine to the colon. In the colon, they can act as preferred fermentation substrates by certain gut-residing probiotic microorganisms (*Bifidobacterium* spp., *Lactobacillus acidophilus,* and *Lactobacillus casei*) which enables these groups of microorganisms to achieve ecological dominance in the human and animal colons (Hidaka *et al*(3)).

TABLE 1: General Properties of Oligosaccharides

* reduced sweetness carbohydrate
* not degraded by enzymes of stomach and small intestine
* utilised by certain probiotic bacteria (bifidus factor)
* modify viscosity and freezing point of food
* affect emulsification, gel formation and gel binding
* has bacteriostatic properties
* alter colour of the food
* act as a humectant and control moisture
* has similar properties to dietary fibre
* has low calorific value
* act as an anti-caries agent

Research into oligosaccharides as functional foods started in Japan in the 1970's. Their commercial production started around 1985. Progress towards the commercial production in Japan of the different oligosaccharides has been reviewed recently (Nakakuki(4)).

This paper reviews the production of oligosaccharides, and differences between particular oligosaccharides. Factors affecting production are discussed. The commercial development of oligosaccharides and their emerging uses in the food industry are summarised.

2. PRODUCTION OF OLIGOSACCHARIDES

Oligosaccharides can be produced chemically or by using enzymes. For food use, oligosaccharides are either extracted from plant or microbial biomass and purified (Okazaki *et al*(5)) or are synthesised enzymically from simple sugars by transglycosylation reactions (Nilsson(6); Prenosil *et al*(7)).

2.1 Extraction from Polysaccharides

In the former cases, polysaccharides are usually extracted from plants and partially hydrolysed to mixtures of oligosaccharides. In some cases, the product is then partially purified by a chemical or physical fractionation process prior to use as a functional food ingredient. Examples of this approach are the processes developed by Suntory Ltd. (Japan) to produce xylo-oligosaccharides, and by Raffinerie Tirlemontoise (Belgium) to produce oligo-fructose and inulo-oligosaccharides. Similarly, neo agar oligosaccharides are being produced form agar (Kono(8)) and malto and iso-malto-oligosaccharides from starch (Nakakuki(9)).

2.2 Synthesis from Sugars

In contrast, synthesis of an oligosaccharide for food use from simple sugars is normally achieved by diverting the normal hydrolytic action of a hydrolase (glycosidase) enzyme, such as β-galactosidase, into transglycolytic activity. In this way, a disaccharide, such as sucrose or lactose, can form oligosaccharides simultaneously with the formation of the normal sugar hydrolytic products.

Hydrolases are glycosidases which can transfer the glycosyl moiety of a substrate to acceptors other than water, and hydrolysis only represents a special case in which water serves as the acceptor(6). Processes developed by Meiji Seika (Japan) to produce fructo-oligosaccharides from sucrose(8), and by Yakult Honsha (Japan) to produce galacto-oligosaccharides from lactose (Matsumomo *et al*(10)) are examples of the current industrial use of glycosidases to synthesize food-grade oligosaccharide mixtures.

2.3 Synthesis using Glycosyltransferases

Oligosaccharides of high specificity can be formed by glycosyltransferases. The glycosyltransferases catalyse the stereo- and regio-specific transfer of a monosaccharide from a donor substrate (a glycosyl nucleotide) to an acceptor substrate. The glycosyltransferase enzymes (in contrast to the situation for glycosidases) occur in low concentrations in nature and are consequently much more expensive to procure than glycosidases. Furthermore, the glycosyltransferases require the use of co-factors (e.g. uridine di phosphate) and are not as useful for formation of shorter-chain oligosaccharides. Their use lies in the formation of specific longer-chain oligosaccharides, and the synthesis of a single oligosaccharide rather than a mixture of oligosaccharides. Synthesis of oligosaccharides by glycosyltransferases has been elegantly reviewed(6).

2.4 Typical Production Process

Production of food-grade oligosaccharides is a relatively simple process, whether it be by an extractive technology or by enzymic synthesis from simple sugars. The process steps are listed in Table 2. Alternative procedures in each step are listed. Unless an immobilised enzyme process can be developed, the cost of purchase of an enzyme for commercial production is prohibitive. Most manufacturers therefore produce by microbial fermentation the required enzymes (e.g. a β-galactosidase) as crude enzyme extracts for use in hydrolysis or transglycosidation. Substrates for the production of oligosaccharide may need to be purified (e.g. de-lignified) or sterilised (e.g. milk powder). If a batch enzymic bioconversion process is used, then the enzyme present in the reacted product has to be deactivated after the appropriate interval to prevent further unwanted hydrolytic reactions taking place. Glycosidases simultaneously hydrolyse and synthesise (transglycosylate). Decolourisation with activated carbon and deionisation of the crude oligosaccharide mixture usually takes place on completion of the enzymic bioconversion. In some processes, these steps also occur prior to the bioconversion as well as after it.

TABLE 2: Typical Process Steps for Oligosaccharide Production

▸	Production of enzyme by microbial fermentation	OR	Purchase enzyme
▸	Crude extraction and purification of polysaccharide	OR	Sterilisation of carbohydrate substrate
▸	Batch enzymic hydrolysis/synthesis then Inactivation of enzyme	OR	Immobilised enzymic bioreactor
▸	Decolourisation and deionisation of crude oligosaccharide mixture		
▸	Fractionate by ion exchange, then Filter, then Spray Dry or Use as Syrup	OR	Use crude oligosaccharide mixture as a food ingredient

The process has to be monitored carefully so that a reproducible mixture of oligosaccharides, hydrolysis products and the substrate carbohydrate is produced. The mixture formed depends on the source of the enzyme used, the reaction time and the initial substrate concentration. Reproducibility of the mixture is important in a commercial situation and this is the most critical aspect of current processes.

2.5 Production of Fructo Oligosaccharides - the Meiji Process

Kono(8) has described the Meiji process to produce the fructo-oligosaccharide mixtures known as Neosugar G, Neosugar P and Meioligo^{R}. Fungal β-fructo-furanosidase (β-FFase) derived from *Aspergillus niger* (ATCC 20611) is used. The enzyme can convert 60 percent of the sucrose substrate to the oligosaccharide. The enzyme derived from *Aureobasidium* spp. is also useful for commercial production (Hayashi *et al*(11). Kono(8) also described the use of an anion exchange resin, Amberlite IRA-94 to immobilise the enzyme. By varying the flow rate of sucrose through the enzyme bioreactor, various mixtures of fructo-oligosaccharides, sucrose, glucose and fructose can be obtained.

Typical sugar proportions in Neosugar formulations for unpurified Neosugar G are; glucose + fructose, 32%; sucrose, 11%; and fructo oligosaccharides 57% (mainly tri- and tetra-oligosaccharides). For the purified Neosugar P (which has been subjected to an anion exchange purification process), the ratios are: glucose + fructose, 1%; sucrose, 4%; fructo oligosaccharides, 95% (Kono, 1993).

A typical sucrose substrate loading for the Meiji process is 50% (w/v) solution and a conversion of 60% of this is achieved.

2.6 Production of Galacto Oligosaccharides - the Yakult Process

A 1992 study by D & K Consulting concludes that the world will continue to have an excess of cheese whey until at least the year 2000 and predicts that 24 to 30 percent of surplus whey will be used during the next 10 years to produce oligosaccharides (Devaux(12)). This translates to a potential world production of 2.25 million tonnes of galacto-oligosaccharides in 1995 rising to 6.74 million tonnes by the year 2000 (Devaux(12)).

If production does achieve these high levels, then this will provide sufficient oligosaccharide for 1500 million people by the year 2000 (based on dosage of 10g/day/person)! Present production capacity by Yakult is believed to be about 7000 tonnes per year, with similar levels produced by other manufacturers.

Promotion of the use of galacto oligosaccharides, rather than, say, fructo oligosaccharides, is based on the finding that oligosaccharides in human milk are responsible for the promotion of growth of bifidobacteria in the colon (a 'bifidus factor') (Mutai and Tanaka(13)). Consequently, Snow Brand Milk Products incorporate galacto oligosaccharides into their infant milk formulae.

Galacto oligosaccharides, comprising mainly di-, tri- and tetra-oligosaccharides, are formed by the action of β-galactosidase on lactose. Galactose is the dominant component sugar in the oligosaccharides. Some penta- and hexa-oligosaccharides are also formed but these are minor components(10) (Smart(14); Toba *et al*(15)).

β-galactosidase (EC 3.2.1.23, β-D-galactosidase galactohydrolase) is also called lactase. This enzyme is widely used to hydrolyse lactose to glucose and galactose industrially, and is an important enzyme in molecular biology as a component of the lac operon, and as a genetic marker. Consequently, it is perhaps the most studied enzyme in molecular biology. This detailed structural knowledge of the enzyme provides an important base for its future industrial development. However, for the present, it is known that β-galactosidases derived from different microbial species vary markedly in their ability to produce oligosaccharides. This variation in the ratios and types of di-, tri- and tetra-oligosaccharides formed during transgalactosylation reactions is also dependent on the lactose concentration used at the start of the reaction, and on the degree of hydrolysis which is simultaneously occurring. All the oligosaccharides being produced, as well as the original lactose are subjected to degradation by hydrolysis to glucose and galactose.

A wide variety of β galactosidases are available commercially from plant, animal and microbial sources. However, only those from microbial sources are available at prices low enough for them to be considered for food grade production of oligosaccharides. For these purposes, β galactosidases from *Kluyveromyces lactis, Bacillus circulans, Aspergillus niger* and *A. oryzae* are available commercially. *Streptococcus thermophilus* has also been used but is not available commercially. Consequently, commercial users frequently produce the required β galactosidase from the appropriate microbial species by fermentation themselves.

The possible reaction products derived from lactose are numerous, and their formation has been discussed(7,14). They are shown in Table 3.

Yakult manufacture a galacto-oligosaccharide mixture called Oligomate 50^R. This is manufactured by reacting a high concentration (40% w/v) of lactose with β galactosidase from *Aspergillus oryzae*. About 30% of the lactose is converted to oligosaccharides. The mixture is then reacted with β galactosidase from *Streptococcus thermophilus* which increases the concentration of oligosaccharides further. The composition of the final solution of Oligomate 50^R is then 50% galacto oligosaccharides, 12% lactose and 38% monosaccharides(10). About one-third of the galacto-oligosaccharides are galacto-disaccharides - the result of using *S. thermophilus*(14). Yakult have a second process which involves anion-exchange separation of mono-, di- and higher-oligosaccharides from each other. This separation is performed after the first stage enzymic bioconversion with *A. oryzae*(10).

TABLE 3: Possible Reaction Products from Lactose

Common Name	Component Linkage	Process
Lactose	Gal (1-4) Glu	-
Galactose	Gal	hydrolysis
Glucose	Glu	hydrolysis
Disaccharides		
	Gal (1-2) Glu	internal rearrangement
	Gal (1-3) Glu	" "
Allolactose	Gal (1-6) Glu	" "
	Gal (1-3) Gal	transgalactosylation
Galactobiose	Gal (1-6) Gal	" "
Trisaccharides		
6'-galactosyl lactose	Gal (1-6) Gal (1-4) Glu	" "
3'-galactosyl lactose	Gal (1-3) Gal (1-4) Glu	" "
	Gal (1-6) Gal (1-6) Glu	internal rearrangement
	Gal (1-6) Gal (1-6) Gal	transgalactosylation
Tetrasaccharides }	similar rearrangements and further	
Pentasaccharides }	transgalactosylations, including some	
Hexasaccharides }	side chain formation	

2.7 Production by Controlled Hydrolysis of Plant Polysaccharides

Suntory Ltd sell two products (Xylooligo 70 and Xylooligo 95) containing different percentages of xylo-oligosaccharides in the mixture. The residual fraction consists of monosaccharides. These products are produced from cotton seed hull bran or from corncob. The raw materials are boiled in alkaline solution, washed and saccharified. Alkali is used because of the solubility of xylan in it (the principle is to solubilise the hemicellulose fraction). Saccharification is performed by endo-1,4-β-xylanase. The enzyme used will liberate side chain sugars as well as cleave xylan. The principal products of the process are xylose, xylobiose and xylotriose, with xylobiose predominating.

Suntory use two processes to purify and concentrate the hydrolysate from xylan. One is a two-stage reverse osmosis process, and the other is ultrafiltration followed by a two-stage reverse osmosis process. The process chosen depends on the raw material used (Sasaki *et al*(16)). Production of xylo oligosaccharides from xylan is an example of a controlled hydrolysis process.

Another example is the production of inulo-oligosaccharides from the chicory plant - a process developed by the Belgian firm, Raffinerie Tirlemontoise. This company produces two products. One is a mixture of long-chain fractions (average DP21) of oligosaccharides derived from the inulin ('Raftiline'), and the other is an oligofructose product ('Raftulose') derived by controlled hydrolysis of the first product. The Raftulose product is composed primarily of 2 to 9 monomers (average DP4).

The tetrasaccharide, raffinose, and the trisaccharide, stachyose, are produced from soybean whey, following removal of protein fractions. The syrup produced from the whey is produced by filtration, decolourisation, desalting and concentration. The resulting mixture contains stachyose (24%), raffinose (8%), sucrose (39%), fructose + glucose (16%), others (13%). This mixture, called soybean oligosaccharides, is produced by Soya Oligo Japan, and used in the drink Oligo CC, by Calpis Food Industry Ltd. (Koga(17)).

Iso-malto-oligosaccharides are produced from starch primarily. The various iso-malto-oligosaccharides comprise a currently important commercial sector of oligosaccharides with production by four Japanese companies exceeding 10,000 tonnes per annum (see Table 4). β-amylase and α-glucosidase are used to saccharify the starch. Products mainly contain isomaltose, panose and isomaltotriaose.

2.8 Commercial Production of Oligosaccharides

Limited information is available on current tonnages produced or sold. The oligosaccharides listed in Table 4 have all been developed at least to the pilot scale and marketing trials conducted. The major oligosaccharides in tonnages are iso-malto-, galacto-, and fructo-oligosaccharides.

3. CURRENT AUSTRALIAN RESEARCH

Procedures to produce galacto-oligosaccharides from lactose using a crude enzyme extract of β-galactosidase from *Streptococcus thermophilus* have been developed. Lactose concentrations of 30 percent were used. Comparisons were made between β-galactosidases from two commercial sources (sources: *Aspergillus oryzae, Kluyveromyces fragilis*) and from a fermentation culture of *S. thermophilus* (Playne *et al*(18)).

Snow Brand Milk Products Ltd. have recently built a major factory in Australia to produce and sell overseas infant milk powders containing galacto oligosaccharides. This plant was commissioned in 1994.

Major research programs are being developed on probiotic bacteria in the CSIRO and at the University of New South Wales in conjunction with several companies. Part of this research is funded by the new Cooperative Research Centre for Food Industry Innovation. The main thrust of this research is to obtain evidence on the health-promoting properties of probiotics. Because of the importance of oligosaccharides in enhancing the growth of bifidobacteria in the colon, these research programs are examining the interaction between oligosaccharides and colonic health.

TABLE 4: Oligosaccharide Production in Japan and Europe

Oligosaccharide	Trade Name	Company	Production 1992 (tonnes)
▸ Fructo-	Mei Oligo/NeoSugar	Meiji Seika Kaisha	4000
▸ Soyabean-	-	Soya Oligo Japan	1000
(raffinose, stachyose)	Oligo CC	Calpis Food Industry	?
	-	Nippon Tensai Seito	?
▸ Isomalto-	ALOM/IMO500	Showa Sangyo	6500
	-	Hayashibara Shoji	2800
	-	Nippon Shito Kohyo	2000
	-	Nippon Shokuhin Kako	800
▸ Galacto-	OligoMate 50/TOS	Yakult Yakuhin Kogyo	6600
	-	Nisshin Sugar	400
	-	Snow Brand Milk Prod.	?
	-	Borculo Whey Products	?
▸ Xylo-	Xylo-Oligo 70/95	Suntory Ltd.	?
▸ Isomaltulose	(Palatinose)	Mitsui Sugar	500
▸ Gentio-	-	Nippon Shokuhin Kako	?
▸ Lactulose	-	Morinaga Milk Industry	?
▸ Malto oligosyl sucrose	(coupling sugar)	Ezaki Glico	?
▸ Malto-	Fuji Oligo 450	Nihon Shokuhin Kako	2000
▸ Lactosucrose	-	Hayashibara Shoji	?
▸ Oligoglucosylsucrose	-	Hayashibara Shoji	?
▸ Cyclodextrins (α, β, δ)		Nihon Shokuhin Kako	?
▸ Branched cyclodextrins	Isoeleat P	Ensiuko	?
▸ Inulo-	Raftiline	Raffinerie Tirlemontoise	?
Oligofructose	Raftilose	" "	?
▸ Gluco-	GOS	BioEurope	?

4. PATENTS ON OLIGOSACCHARIDES

Over 90 percent of the patent activity on oligosaccharides is in Japan. The earliest patent on a process to produce a food-grade oligosaccharide is that of Mutai *et al* assigned to Yakult Honsha. The application date was 7 August 1980. A search over the period 1980 to 1988 revealed 36 production patents. Patent activity has accelerated remarkably since 1989 with some 260 patents over 4.5 years. In addition to patents on methods of production, the last five years have seen many patents on new applications for oligosaccharides, new methods of analysis, and methods to produce linked-oligosaccharides.

Of the 260 patents since 1989, the distribution of activity has been for malto- and iso-malto- (16%), galacto- (15%), inulo- and fructo- (13%), xylo- (5%), chito- (5%) and gluco-oligosaccharides (5%). The remaining 41% is not specific to a particular class of oligosaccharide.

Patent applications have been assigned to a wide range of companies with the most activity by Yakult Honsha, Showa Sangyo, Meiji Seika Kaisha, Snow Brand Milk Products, Nisshin Sugar, Mitsui Sugar, Nihon Shokuhin Kako and Calpis Food Industry. Non-Japanese companies have shown remarkably little activity in this area.

5. USES OF OLIGOSACCHARIDES

New uses for oligosaccharides suggested by patents are dominantly as sweeteners and as bifidus factors promoting the growth of intestinal bacteria. Other applications include: fatty acid esters of fructose oligosaccharide as an emulsifying and gelling agent in cosmetics; a lozenge and mouth rinse for control of oral bacteria using lacto difucotetraose; isomalto-oligosaccharide as an immunostimulant; chitin oligosaccharide in cosmetics; oligosaccharides added to candies; inulin oligosaccharide as a plant growth promoter; inhibition of *Salmonella* in animals with fructo oligosaccharide; use in chewing gums; galacto-oligosaccharide being used to promote calcium absorption in the gut; and galacto-oligosaccharide being incorporated into infant milk powder formulae; and sulphated oligosaccharides being used as antiviral agents. However, the major use remains incorporation of oligosaccharides into drinks and sweets, where the oligosaccharide acts as a sweetener and as a 'bifidus' factor. Examples are Calpis' Oligo CC and Meiji's candies.

6. ACKNOWLEDGMENTS

I am thankful to my colleagues, Angela Dimopoulos, Bernadette Morel, Katherine Scurrah and Azra Pachenari for their excellent work and for discussions. The financial contributions made by CSIRO, the Dairy Research and Development Corporation, United Milk Tasmania Ltd. and the CRC for Food Industry Innovation are gratefully acknowledged. I thank the Organising Committee of this International Symposium on Bioproducts Processing for inviting me and for meeting expenses.

7. REFERENCES

1. Van den Broek, A., 1993, Internat.Food Ingred., No1/2, 4-9.

2. Rademacher, T.W., Parekh, R.B. and Dwek, R.A., 1988, Ann.Rev.Biochem., 57, 785-838.

3. Hidaka, H., Eida, T., Takizawa, T., Tokunaga, T., and Tashiro, Y., 1986, Bifidobacteria Microflora 5, 37-50.

4. Nakakuki, T. (Editor), 1993, "Oligosaccharides" Japanese Technology Reviews, vol 3, number 2, 235 pages (Gordon and Breach Science Publishers).

5. Okazaki, M., Fujikawa, S. and Matsumomo, N., 1990, Bifidobacteria Microflora 9, 77-86.

6. Nilsson, K.G.I., 1988, Trends Biotechnol. 6, 256-264.

7. Prenosil, J.E., Stucker, E. and Bourne, J.R., 1987, Biotechnol.Bioeng. 30, 1019-1025.

8. Kono, T., 1993, Japanese Technology Reviews 3(2), 204-217.

9. Nakakuki, T., 1993, Japanese Technology Reviews 3(2), 1-24.

10. Matsumoto, K., Kobayashi, Y., Ueyama, S., Watanabe, T., Tanaka, R., Kan, T., Kuroda, A. and Sumihara, Y., 1993, Japanese Technology Reviews 3(2), 90-106.

11. Hayashi, S., Nonokuchi, M., Imada, K. and Ueno, H., 1990, J.Indust.Microbiol. 5, 395-400.

12. Devaux, W., 1993, Internat.Food Ingred. No1/2, 39-44.

13. Mutai, M. and Tanaka, R., 1987, Bifidobacteria Microflora 6, 33-41.

14. Smart, J.B., 1993, Bull.Internat.Dairy Fed. No 289, 16-22.

15. Toba, T., Tomita, Y., Itoh, T. and Adachi, S., 1981, J.Dairy Sci. 64, 185-192.

16. Sasaki, H., Shiba, H. and Matsumoto, N., 1993, Proc.6th Europ.Congr.Biotechnol.,Firenze (Poster).

17. Koga, Y., Shibata, T. and O'Brien, R., 1993, Japanese Technology Reviews 3(2), 175-202.

18. Playne, M.J., Morel, B. and Dimopoulos, A., 1993, Proc.11th AustralianBiotechnol.Conf., Perth, p207

GAS HOLDUP CORRELATION FOR AERATED STIRRED VESSELS

R. Parthasarathy, N. Ahmed and G.J. Jameson
Dept. of Chemical Engineering, University of Newcastle, N.S.W., Australia.

The generation of gas holdup in aerated stirred vessels was studied experimentally as a function of the impeller speed and bubble size. Bubbles of various sizes were generated, independently of the impeller, and their mean size and the gas holdup were observed as functions of the impeller speed. Experimental results obtained for various impellers show the roles of both the specific power and bubble size in determining gas holdup. A quantitative measure of the increase in holdup with decreasing bubble size, all other factors being similar, is demonstrated. A dimensionless correlation based on fluid mechanical behaviour of two-phase dispersion in stirred vessels is proposed for estimating the gas holdup.

1. INTRODUCTION

Gas holdup is one of the important design parameters in aerobic biochemical stirred reactors and influences the transport processes between the gas phase and the fermentation broth. Gas holdup has been studied as a function of important operating variables like impeller speed and superficial gas velocity and numerous correlations, based on detailed experimental investigations, have been proposed for estimating the gas holdup in stirred gas-liquid contactors. A list of some of the suggested correlations is presented in Table 1. Most of these correlations are empirical without any fundamental fluid mechanical basis and, therefore, contribute very little to the understanding of the gas holdup characteristics. In more recent studies, attempts have been made to propose holdup correlations based on different impeller and bulk flow regimes [Warmoeskerken (13) and Smith (14)]. In summary, in most of the reported correlations, the specific power input or the impeller speed and the gas flow rate are the operating parameters that have been considered to influence the gas holdup. The effect of bubble size, which is one of the important hydrodynamic variables, on gas holdup has not yet been investigated in detail. Calderbank (2) incorporated the effect of bubble size in his correlation in terms of bubble terminal rise velocity. However, the bubble size was considered to remain constant over the range of operating conditions used. The lack of information on bubble size and its effect on gas holdup in stirred vessels is understandable considering the difficulties involved in measuring bubble size.

The present study is an attempt at observing the effect of bubble size on gas holdup in aerated stirred vessels. Bubbles of various initial sizes were generated using different spargers, and bubble breakup and its effect on the holdup were monitored as a function of the impeller speed. It was possible to study the breakup process in isolation by maintaining a non-coalescing environment. A dimensionless correlation based on two-phase hydrodynamics in stirred vessels is proposed for estimating the gas holdup.

Table 1. Literature correlations on gas holdup

Author(s)	System	Impeller	T, (m)	Correlation
Foust *et al* (1)	water	6- to 10-curved bladed turbines	0.32 2.44	$\phi_g = \text{constant}\left[\frac{P_g}{V}\right]^{0.47} U_g^{0.53}$
Calderbank (2)	water glycerol EtOH EtAc	6-bladed Rushton turbine	0.20 0.50	$\phi_g = 2.16 \times 10^{-4}\left[\frac{(P_g / V)^{0.4} \rho_L{}^{0.2}}{\sigma^{0.4}}\right] \left[\frac{U_g}{U_t}\right]^{0.5} + \left[\frac{U_g \phi_g}{U_t}\right]^{0.5}$
Yoshida and Miura (3)	water NaOH glycerol	12-bladed Rushton turbine	0.25 0.59	$\phi_g = \text{constant}\ N^{0.8} U_g{}^{0.75} D^{1.2}$
Rushton and Bimbinet (4)	water, corn syrup	6-bladed Rushton turbine	0.23 0.91	$\frac{\phi_g}{1-\phi_g} = 0.31\left(\frac{P_g}{V}\right)^{0.31} U_g{}^{0.60}$
van Dierendonck *et al* (5)	water, organic liquids	6-bladed Rushton turbine	0.17 0.45	$\phi_g = 0.31\left(\frac{\mu_L U_g}{\sigma}\right)^{2/3}\left(\frac{\rho_L \sigma^3}{g \mu_L{}^4}\right)^{1/6} + 0.45\left(\frac{N - N_o{}^*}{T(gT)^{1/2}}\right) D^2$
Miller (6)	various liquids	4-flat-bladed paddle	0.15 0.69	$\phi_g = 2.16 \times 10^{-4}\left[\frac{(P_e / V)^{0.4} \rho_L{}^{0.2}}{\sigma^{0.4}}\right] \left[\frac{U_g}{U_g + U_t}\right]^{0.5} + \left[\frac{U_g \phi_g}{U_g + U_t}\right]^{0.5}$
Sterbacek and Sachova (7)	water glycerol	4-bladed Rushton turbine	0.16	$\phi_g = 9.0\text{x}10^{-5}\left(\frac{\rho_L U_g D}{\mu_L}\right)^{0.77}\left(\frac{\rho_L N^2 D^3}{\sigma}\right)^{0.5}$
Loiseau *et al* (8)	organic and ionic solutions	6-bladed Rushton turbine	0.22	$\phi_g = 0.011\left(\frac{P_e}{V}\right)^{0.27}\left(\frac{U_g}{\sigma}\right)^{0.36} \mu_L{}^{-0.056}$

Hassan and Robinson (9)	water and non-ionic solutions	6-bladed Rushton turbine	0.15 0.29	$\phi_g = 0.113\left(\frac{QN^2}{\sigma}\right)^{0.57}$
Yung *et al* (10)	water glycol acetone NaCl Na_2SO_4	6-bladed Rushton turbine	0.4	$\phi_g = 0.52\left(\frac{Q}{ND^3}\right)^{0.5}\left(\frac{\rho_L N^2 D^3}{\sigma}\right)^{0.65}\left(\frac{D}{T}\right)^{1.4}$ $\phi_g = \text{constant } Q^{0.5}N^{0.8}D^{1.4}$
Hughmark (11)	water cyclo-hexane	6-bladed Rushton turbine	0.17 0.46	$\phi_g = 0.74\left(\frac{N^2T^4}{gDV^{2/3}}\right)^{0.5}\left(\frac{d_b N^2 T^4}{\sigma V^{2/3}}\right)\left(\frac{Q}{NV}\right)^{0.5}$
Chapman *et al* (12)	water	various types	0.56	$\phi_g = 17.9\left(\frac{P_g}{V}\right)^{0.31} U_g^{0.67}$
Warmoes-kerken (13)	water	various types	0.44	$\phi_g = 0.133Q^{0.35}ND^{0.16}$ (Before 3 - 3 structure) $\phi_g = 0.177Q^{0.4}N^{0.9}D^{0.1}$ (After 3 - 3 Structure)
Smith (14)	low-viscosity liquids	various tyes	----	$\phi_g = 0.85\left(\frac{N^2 Q \rho_L}{g\,\mu_L}\right)^{0.35}\left(\frac{D}{T}\right)^{1.25}$

2. EXPERIMENTAL

Experiments were carried out in two standard baffled perspex tanks of diameters (T) 0.195 m and 0.4 m. The liquid height (H) was maintained equal to the tank diameter. Four different impellers were used in this study, namely, the Rushton turbine, flat blade impeller, downward-pumping 45° pitched-blade disc turbine and marine propeller. Whereas the flat blade impeller and the marine propeller were not used in the 0.4 m tank all four impellers were used in the 0.195 m tank. The impeller diameter (D) was equal to T/3 and its clearance from the tank bottom was equal to its diameter.

Three spargers were used to generate bubbles of different initial sizes. They were a ring sparger and two sintered glass discs of porosities 150-250 μm and 5-15 μm, designated as porosity 0 and porosity 4, respectively, by the manufacturer (Pyrex). In the following discussion, the sintered glass discs will be denoted by their respective classification numbers, i.e. sparger 0 and sparger 4. The sintered glass discs used in the 0.195 m and 0.4 m tanks were 90 mm and 65 mm in diameter, respectively. The discs were fixed to the tank bottom directly underneath the impeller in both tanks. A single disc was used in the 0.195 m tank whereas four discs were used in the 0.4 m tank. The ring sparger was located halfway between the impeller and the tank bottom. Sparger 0 was not used in the 0.4 m tank whereas all three spargers were used in the 0.195 m tank.

Bubble size was measured by photography. Special cells, similar in design to the one used by Ahmed and Jameson (15), were used for photographing the bubbles. A sample of liquid containing the bubbles was drawn continuously from the tank, passing between two transparent plates either 2 or 6 mm apart, where the bubbles could be photographed by rear illumination. The negatives of the photographs were projected onto a digitiser pad and the bubbles were sized using a microcomputer. A 2 mm graticule pasted to the inside wall of the rear window of the photographic cell was used as the calibration scale. For each experimental condition, a sufficient number of bubbles was sized to yield statistically significant results. The bubble size was measured at the mid-liquid height position in the tank.

Air and water were used as the gas and liquid phases respectively. To minimize bubble coalescence, methyl isobutyl carbinol (MIBC) was added at a concentration of 50 ppm V/V. The surface tension of the liquid medium was found to be 71.3 mN/m. The noncoalescing nature of the medium was confirmed by measuring the bubble size in the vicinity of the impeller and at various locations in the bulk of the tank. A statistical analysis confirmed that the sizes of the bubbles coming out of the impeller zone were preserved in other regions of the tank.

The gas holdup (ϕ_g) was determined by measuring the change in the pressure difference between two pressure tappings in the tank wall when the gas was sparged. A differential pressure transmitter was used for this purpose. The gas holdup values were corrected for the dynamic pressure effects experienced by the differential pressure transmitter due to the turbulent liquid flow generated by the impeller agitation.

The power draw by the impeller was determined by measuring the torque experienced by the impeller-motor assembly unit using a load cell. However, details of impeller power consumption are not discussed in this paper. The impeller speed (N) was varied between 2.08 rps and 16.67 rps (125 rpm to 1000 rpm) and the superficial gas velocity (U_g) was varied between 2.5×10^{-4} m/s and 1.25×10^{-3} m/s.

3. RESULTS AND DISCUSSION

3.1 Effect of Impeller Speed on Mean Bubble Size

In the present study, Sauter mean bubble diameter (d_{32}), which is a commonly used mean size representative of the bubble size distribution, was used in the analysis. The d_{32} value measured at an impeller speed of 2.08 rps (125 rpm) is designated as the initial bubble size. This is the minimum impeller speed at which bubbles generated by the spargers were sufficiently dispersed radially for the bubble size measuring technique devised for this work to be utilized.

With increase in impeller speed, the d_{32} values of the bubbles generated by the three spargers followed a certain trend, consistent for all the impellers used in both tanks. As an illustration, the d_{32} values obtained in the 0.195 m tank using the Rushton turbine are shown in Figure 1. The initial d_{32} values were found to be 300 μm, 2.5 mm and 5 mm for sparger 4, sparger 0 and ring sparger, respectively. With increase in impeller speed, the 300 μm bubbles were unaffected and their size remained nearly constant. On the other hand, the d_{32} value of the ring sparger bubbles decreased continuously with increase in impeller speed. The d_{32} value of sparger 0 bubbles also decreased with increase in impeller speed but only after 5.0 rps. Similar trends in d_{32} values were observed with increase in agitation for other impellers used in both tanks. The actual d_{32} values and the initiation of the breakup, however, varied with impeller type and scale, obviously related to the specific power input. The breakup behaviour of bubbles of different initial sizes has been discussed in detail by Parthasarathy and Ahmed (16), (17). Superficial gas velocity, in the range used, has no significant influence on the trend observed in the bubble size results. Thus, two categories of bubble size were observed in the present study: 300 μm bubbles, the size of which remained practically constant (no breakup) over the range of operating conditions used and, the bubbles generated by sparger 0 and ring spargers which underwent breakup with increase in agitation. The situation makes it possible to evaluate independently the generation of gas holdup as functions of both the impeller characteristics and the bubble size.

3.2 Effect of Impeller Speed on Gas Holdup

The gas holdup results obtained in the 0.195 m tank using the bubbles generated by the three spargers are shown in Figure 2 for the Rushton turbine. It can be seen that both impeller speed and bubble size influence the gas holdup values. At low impeller speeds, the 300 μm bubbles led to higher gas holdup when compared to those obtained with sparger 0 and ring sparger bubbles. But as the impeller speed increased, the difference in the holdup values narrowed which is clearly due to the breakup and consequent decrease in the size of sparger 0 and ring sparger bubbles. Similar trends were observed in the gas holdup results obtained with the other impellers used in both tanks. However, the magnitude of gas holdup values obviously varied with the impeller type and the size (scale) of the tank.

To illustrate the effect of impeller type, the gas holdup values obtained using the pitched-blade disc turbine in the 0.195 m tank (Figure 3) are compared to those obtained with the Rushton turbine (Figure 2). At almost all impeller speeds, the pitched-blade disc turbine generated lower gas holdup for all three spargers. The relative superiority of the Rushton turbine in generating higher gas holdup can be appreciated by comparing the gas holdup values for the 300 μm bubbles. Since the mean bubble size remained constant, gas holdup was determined mainly by the dispersion ability of the impeller. In case of sparger 0 and the ring sparger bubbles, the gas holdup achieved was not only due to the impeller pumping but also to the decrease in bubble size.

The gas holdup values obtained using the Rushton turbine in the 0.4 m tank are shown in Figure 4. By comparing these results with those obtained in the 0.195 m tank (Figure 2), the effect of scale can be gauged. Although the gas holdup was nearly the same in both tanks at low impeller speeds, it was higher in the 0.4 m tank at higher impeller speeds. As the scale was increased by a factor of two, the power consumption of the impellers for the same impeller speed was increased by a factor of 32 (=2^5) according to the impeller power number relationship

$$Np = \frac{P}{\rho_L N^3 D^5} \qquad (1)$$

where N_p is the impeller power number, ρ_L is the continuous phase density, and P, N, D are impeller power consumption, speed and diameter, respectively. Hence, for the same speed, the impeller employed in the 0.4 m tank had a higher specific power input. As bubble size and impeller pumping to some extent are functions of specific power input, we may conclude that the impeller employed in the larger tank produced smaller bubbles and therefore, higher gas holdup. It should be indicated at this point that the enhancement in gas holdup due to the different scales commenced only when the specific power in the larger tank exceeded the highest specific power in the smaller tank.

3.3 Correlation for Gas Holdup

The prediction of gas holdup from basic principles is difficult because the necessary information on single- and two-phase flow in stirred vessels is lacking significantly. Hence, the emphasis is generally on developing predictive correlations. Many of the correlations reported in the literature (Table 1) are based on specific power input (P_g/V) and superficial gas velocity (U_g). These correlations are mostly dimensional and do not represent two-phase flow behaviour in stirred vessels adequately. Hence their application is limited to the system and the scale for which they were developed.

Calderbank (2) considered the effect of two-phase flow on stirred vessel hydrodynamic variables and proposed the semi-empirical equation for gas holdup that is shown in Table 1. This equation was obtained using the concept of critical Weber number as applied to bubble breakup in turbulent flow. The correlation accounts for the effect of bubble size on gas holdup by incorporating the bubble terminal rise velocity U_t. Moreover, it predicts a finite gas holdup at zero specific impeller power input which is equivalent to the gas holdup for bubble column under bubbly flow conditions. Miller (6) proposed a gas holdup correlation by modifying Calderbank's equation and using a slightly different definition of gas holdup.

Although Calderbank's correlation has been used widely in predicting the gas holdup in stirred vessels, it should be pointed out that it relies on information on impeller specific power input which itself is a dependent variable. Moreover, in obtaining the correlation, Calderbank varied the bubble size only over a narrow range and assumed that bubble terminal rise velocity (U_t) is constant. This assumption may be valid for bubble sizes of 2 mm or above but will prove to be erroneous for much smaller bubbles experienced in industrial reactors, especially at very high specific power inputs.

The gas holdup in stirred vessels is influenced by many factors but mainly by impeller speed, gas flow rate and bubble size. In general, gas holdup increases with increase in gas flow rate due to an increase in the number of bubbles in the system at steady state. The increase in gas holdup with increase in impeller speed results from the enhanced impeller pumping capacity and the breakup of the larger bubbles into smaller ones. A decrease in bubble size leads to higher gas holdup because of the lower terminal rise velocities of smaller bubbles, leading to longer gas phase residence time. The impeller pumping capacity (Q_p) is related to the impeller speed (N) and diameter (D) by the following relationship :

$$Q_P = N_{QP} N D^3 \tag{2}$$

where N_{QP} is the impeller pumping number. The pumping number is a function of impeller type and geometry. The differences in gas holdup values obtained for the same impeller speed with different impellers may be attributed partly to the different pumping capacities. Thus, the gas holdup achieved in stirred vessels is a result of the interaction between the gas flow rate, the impeller speed and the bubble size.

Two dimensionless groups can be written to represent the interaction between the variables that determine gas holdup in stirred vessels. They are: (U_g/U_t) and ($N_{QP}ND^2/TU_t$). The first group (U_g/U_t) is the ratio of the superficial gas velocity and the bubble terminal rise velocity, and represents the gas holdup that is due to the gas sparging alone. The second group represents the ratio of the average liquid circulation velocity in the tank ($\propto ND^2/T$) and the bubble terminal rise velocity U_t. An estimate of the average liquid velocity is obtained customarily by dividing the liquid pumping rate of the impeller Q_p by the liquid volume in the tank. The second group may also be obtained by dividing the average liquid circulation time by the bubble rise time, and simplifying the resulting expression. Thus, this group represents the gas holdup fraction that results from the interaction between the liquid flow generated by the impeller rotation and the motion of bubbles coursing through the system.

Using these two dimensionless groups and Calderbank's approach, the following correlation is proposed for predicting the gas holdup in stirred vessels:

$$\phi_g = \left[\frac{U_g \phi_g}{U_t}\right]^{0.5} + A\left[\frac{N_{QP} N D^2}{T U_t}\right]^B \left[\frac{U_g}{U_t}\right]^C \qquad (3)$$

where A, B and C are constants that are to be determined using the experimental values of ϕ_g, U_g, N and U_t.

The bubble terminal rise velocities for the experimentally measured d_{32} values have been determined using the method of Clift *et al* (18). Another parameter whose value has to be specified in order to determine the values of the constants in equation (3) is the impeller pumping number (N_{QP}). A considerable number of studies has been reported in the literature on the pumping capacities of various impellers. However, most of these studies have been limited to the liquid phase. Under aerated conditions, especially at high gas flow rates, the pumping capacity of the impeller drops considerably because it is enveloped completely by the gas. In the absence of reliable information on pumping capacity under aerated conditions, the impeller pumping numbers for unaerated conditions are chosen for the present analysis. The error involved in such calculation is expected to be small since the gas flow rates used in this work were quite low. The N_{QP} values used in the analysis are as follows: 1.3 for the Rushton turbine (Holmes *et al,* 19), 0.69 for the flat blade impeller (Sano and Usui, 20), 0.55 for the propeller (Porcelli and Marr, 21), and 0.92 for the pitched-blade disc turbine (Medek and Fort, 22).

The values of the unknown parameters in equation (3) were determined using a non-linear least squares regression analysis and are shown in Table 2. The values of the correlation coefficient (R) obtained in the analysis are also shown in the table. The regression analysis for the 300 μm bubbles was performed separately because the bubble size effect on gas holdup in this case was constant.

The gas holdup values predicted using the correlations are compared with the experimental values in Figures 5 and 6 for the 300 μm bubbles, and the ring sparger and sparger 0 bubbles respectively. The agreement between the predicted and experimental values is found to be satisfactory except for a few data at high gas holdup values.

The proposed correlation is dimensionless and based on independent variables. It predicts a finite gas holdup when the impeller speed is zero. For using this correlation, an estimation of mean bubble size is required as a function of impeller speed or specific power input. Bubble size equations proposed by Calderbank (2) or Parthasarathy *et al* (23) could be used to estimate d_{32} values approximately. Indications are that the correlation for bubbles undergoing breakup with increase in agitation would not be applicable at diameters below approximately 400 μm. To predict the gas holdup for such small bubbles, the correlation for 300 μm bubbles would be recommended and would probably be applicable for systems with very high specific power inputs, especially if the medium were noncoalescing.

4. CONCLUSIONS

By generating bubbles of various initial sizes, independently of the agitation, it has been possible to observe experimentally the role of bubble size on the gas holdup as a function of the impeller design and rotational speed. It has been demonstrated that, all other factors being similar, smaller

Table 2. Correlation parameters for equation (3).

	A	B	C	R
300 μm bubbles	1.3×10^{-4}	1.72	-0.11	0.82
Ring sparger and sparger 0 bubbles	0.021	0.93	0.51	0.8

bubbles generate higher gas holdup. Gas holdup however, is ultimately determined by the specific power input (breakup and production of finer bubbles) and the liquid flow pattern (circulation of the bubbles) generated by the impeller. Since the use of high specific power is not always permissible in biochemical reactors due to the shear sensitivity of the medium, the practicality of introducing externally generated bubbles into such reactors, to generate larger holdups (transfer rates), should be evaluated.

A dimensionless correlation based on stirred vessel hydrodynamics has been proposed using the experimental data for predicting the gas holdup. The effect of bubble size has been incorporated in the correlation by including the bubble terminal rise velocity. The gas holdup values estimated by the correlation agree well with the experimental values.

5. SYMBOLS

C impeller clearance from tank bottom, m
D impeller diameter, m
d_{32} Sauter-mean bubble diameter, m or μm
d_b bubble size, m or μm
g gravitational constant, m/s^2
H liquid height, m
N impeller speed, rps
N_o minimum impeller speed for gas dispersion, rps
P impeller power consumption, W
P_e power input by sparging plus impeller power input, W
P_g impeller power consumption under aerated conditions, W
Q gas flow rate, m^3/s
R correlation coefficient
T tank diameter, m
U_g superficial gas velocity, m/s or mm/s
U_t bubble terminal rise velocity, m/s
V liquid volume in the tank, m^3

Greek symbols
μ_L liquid viscosity, Pa.s
ρ_L liquid density, kg/m^3
σ surface tension, N/m
ϕ_g gas holdup

6. REFERENCES

1. Foust, H.C., Mack, D.E. and Rushton, J.H., 1944, Ind. Eng. Chem., 36, 517-522.
2. Calderbank, P.H., 1958, Trans. I. Chem. E., 36, 443-463.
3. Yoshida, Y. and Miura, Y., 1963, Ind. Eng. Chem. Process. Dev., 2, 263-268.
4. Rushton, J.H. and Bimbinet, J.J., 1968, Can. J. Chem. Eng., 46, 16-21.
5. van Direndonck, L.L., Fortuin, M.H. and Venderbos, D., 1968, Proc. 4th Eur. Symp. on Chem. Reaction Eng., Brussels, Belgium, pp. 205-215.
6. Miller, D.N., 1974, A. I. Ch. E. J., 20, 445-443.
7. Sterbacek, Z. and Sachova, M., 1976, Int. Chem. Eng., 16, 104-109.
8. Loiseau, B., Midoux, N. and Charpentier, J.C., 1977, A. I. Ch. E. J., 23, 931-935.
9. Hassan, I.T.M. and Robinson, C.W., 1977, A. I. Ch. E. J., 23, 48-56.
10. Yung, C.N., Wong., C.W. and Chang, C.L., 1979, Can. J. Chem. Eng., 57, 672-676.
11. Hughmark, G.A., 1980, Ind. Eng. Chem. Process Des. Dev., 19, 638-641.
12. Chapman, C.M., Nienow, A.W., Cooke, M. and Middleton, I.C., 1983, Chem. Eng. Res. Des., 61, 82-95.
13. Warmoeskerken, M.M.C.G., 1986, *Gas-liquid dispersing characteristics of turbine agitators*, Ph.D thesis, Delft University of Technology, Netherlands.
14. Smith, J.M., 1991, Proc. 7th Eur. Cong. on Mixing, Brugge, Belgium, pp. 233-241.
15. Ahmed, N. and Jameson, G.J., 1985, Int. J. Miner. Process., 14, 195-215.
16. Parthasarathy, R. and Ahmed, N., 1993, A.I.Ch.E. Symp. Series, No.293, Vol. 89., 97-104.
17. Parthasarathy, R. and Ahmed, N., 1994, Ind. Eng. Chem. Res., 33, 703-711.
18. Clift, R., Grace, J.R. and Weber, M.E., 1978, *Bubbles, Drops and Particles*, Academic Press, New York.
19. Holmes, D.B., Vonken, R.M. and Dekker, J.A., 1964, Chem. Eng. Sci., 19, 201-208.
20. Sano, Y. and Usui, H., 1985, J. Chem. Eng. Japan, 18, 47-52.
21. Porcelli, J.V. and Marr, G.R., 1962, Ind. Eng. Chem. Fund., 1, 172-179.
22. Medek, J. and Fort, I., 1979, Proc. 3rd Eur. Conf. on Mixing, Vol.2, Univ. of York, York, England, pp 1-10.
23. Parthasarathy, R. Jameson, G.J. and Ahmed, N., 1991, Chem. Eng. Res. Des., 69, 295-301.

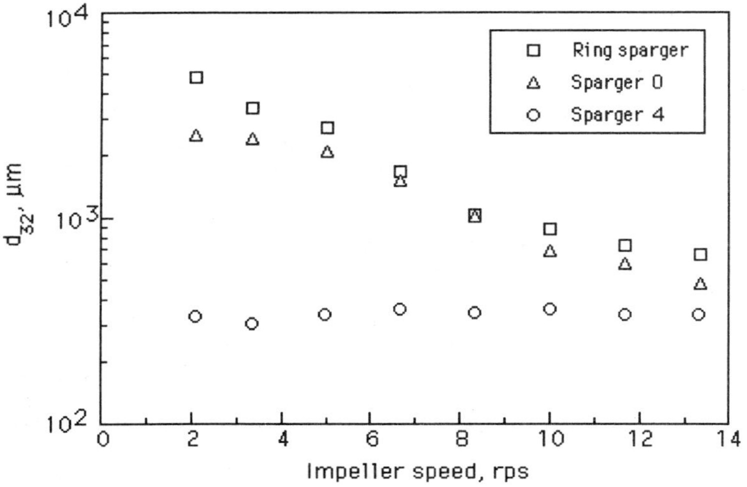

Figure 1. Effect of impeller speed on Sauter-mean bubble diameter, 0.195-m tank, Rushton turbine, $U_g = 1.25 \times 10^{-3}$ m/s.

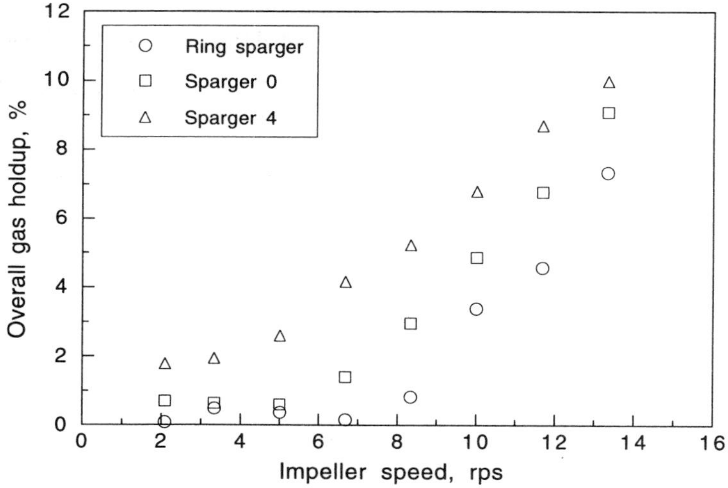

Figure 2. Effect of impeller speed on gas holdup for the bubbles obtained with three different spargers, 0.195-m tank, Rushton turbine, $U_g = 1.25 \times 10^{-3}$ m/s.

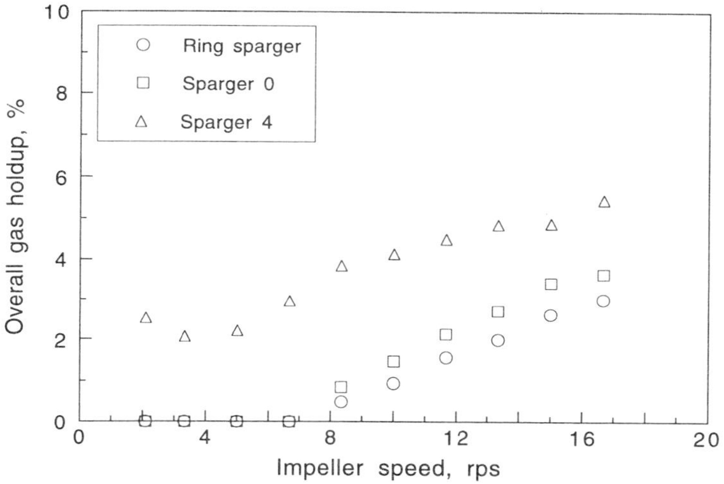

Figure 3. Effect of impeller speed on gas holdup for the bubbles obtained with three different spargers. 0.195-m tank, Pitched-blade disc turbine, $U_g = 1.25 \times 10^{-3}$ m/s.

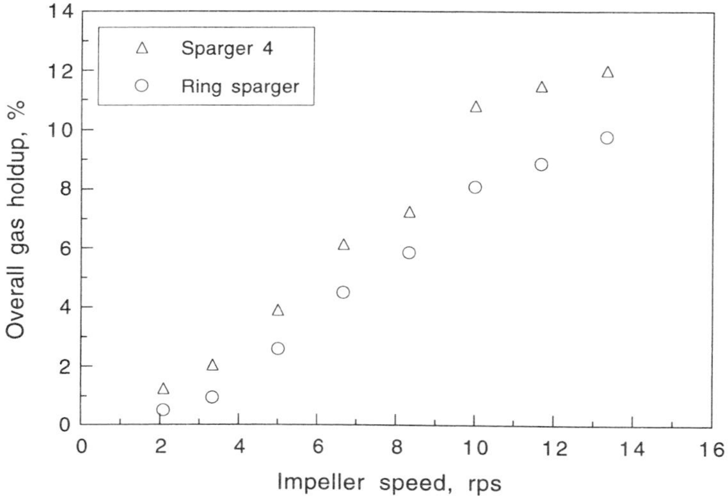

Figure 4. Effect of impeller speed on gas holdup for the bubbles obtained with two different spargers, 0.4-m tank, Rushton turbine, $U_g = 1.25 \times 10^{-3}$ m/s.

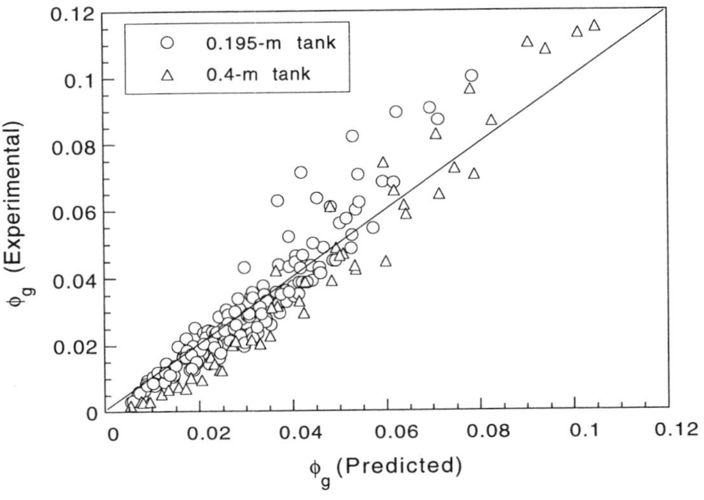

Figure 5. Comparison of experimental and predicted gas holdup values, 0.195- and 0.4-m tanks, 300-μm bubbles, all impellers.

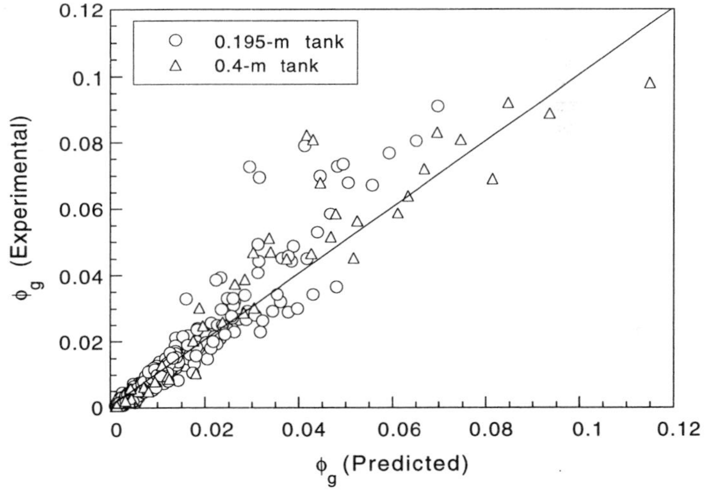

Figure 6. Comparison of experimental and predicted gas holdup values, 0.195- and 0.4-m tanks, all impellers, sparger 0 and ring sparger bubbles.

ENGINEERING AND MICROBIOLOGICAL ASPECTS OF THE PRODUCTION OF MICROBIAL POLYSACCHARIDES: XANTHAN AS A MODEL

Enrique Galindo
Institute of Biotechnology, National University of Mexico, Apdo. Post. 510-3, Cuernavaca, Morelos, 62271, MEXICO.

Xanthan gum is a widely used microbial polysaccharide and its production process is a challenging model of study from the microbiological and from the engineering points of view. This paper reviews the work done at the author's laboratory in the following topics: screening of *Xanthomonas* strains, preservation methods for the bacteria, rheological characterization of broths, mixing problems during fermentation and recovery of the gum by precipitation.

1. INTRODUCTION

Microbial polysaccharides are becoming more important as new materials and ingredients for a variety of applications [Galindo (1), Sutherland (2)]. Xanthan, dextran, pullullan and gellan are industrially produced and a number of a new gums are emerging as potentially important. Currently, xanthan gum is -commercially speaking- the most important microbial polysaccharide. Worldwide annual production of xanthan is about 25,000 tons.

Xanthan fermentation has been also an excellent model of study. Broth viscosity undergoes changes of about four orders of magnitude during culture and broth rheology involves shear thinning, yield stress and viscoelastic characteristics. This makes mixing a key parameter in xanthan fermentation, determining its productivity.

Xanthan is produced by the bacterium *Xanthomonas campestris*, a phytopatogen which infects mainly plants of the family *Cruciferae.* The isolation and screening of strains from natural habitats is a tool which can lead to the selection of strains producing more xanthan or showing novel rheological properties.

Bacteria preservation is a motive of concern as high reproducibility is required in any industrial fermentation process. We have assessed methods involving the successive transfers in "slopes" and the use of cabbage seeds as means of culture preservation.

Xanthan recovery plays an important role in the economics of the process. Up to 50% of the total costs are linked to the downstream operations. We have studied the precipitation of xanthan broths using isopropanol in a stirred tank with a variety of impellers types.

Figure 1 shows a schematic diagram of a typical process

for the production of xanthan gum. Figure 1 also indicates the aspects that have been studied by the author's group and collaborators, namely: bacteria screening and preservation, broth rheology, mixing, and downstream aspects such as heat treatment and xanthan recovery by precipitation. These aspects will be outlined in this paper.

2. MATERIALS AND METHODS

Methodologies have been described in previous papers, as detailed as follows in terms of the subject: phytopatogenicity [Ramírez *et al* (3)], strain screening [Torrestiana *et al* (4), Torres *et al* (5), Ramírez (6), Galindo *et al* (7)], preservation methods [Salcedo *et al* (8), Galindo *et al* (9)], broth rheology [Torres *et al* (5), Galindo *et al* (10), Hannote *et al* (11), Torres *et al* (12)], mixing [Galindo *et al* (13), Xue-ming *et al* (14), Torrestiana *et al* (15), Galindo and Nienow (16), Sánchez *et al* (17), Galindo and Nienow (18)] and precipitation [(Albiter *et al* (19)].

3. RESULTS AND DISCUSSION

3.1 Strain screening

As there is increased evidence that no correlation exists between *Xanthomonas* morphological characteristics and their ability to produce xanthan, alternative methods of selection are necessary. We have developed an improved productivity test (7) using well-mixed 500 ml baffled shake flasks. These flasks were compared with 2800 ml Fernbach flasks. We showed that bacterial growth rates were similar in both types of flasks although the Fernbach flasks gave higher biomass concentrations. Xanthan production was similar in both types of flasks but different viscosities were attained. One a weight basis, the xanthan produced in baffled flasks was up to three times more viscous and more shear thinning. This technique is suitable to screen large numbers of isolates in a minimum of space and allows to obtain xanthan of rheological quality similar to that obtained in fermenters.

Culture virulence can be also used as a criterion for strain selection. We have shown (3) (figure 2) that virulence correlated with xanthan production. As a result of an extensive screening program (6) carried out in the central part of México, a number of potential strains have been isolated. Figure 3 shows typical results regarding the evaluation of various isolates. Some of these strains have been further characterized and improved strains (if compared with a collection strain) have been obtained. One of these strains is able to produce 20 % more xanthan than its collection counterpart (9) and turned out to be comparable, in terms of rheological quality, to a xanthan from a commercial source (5).

3.2 Bacteria preservation

We have shown that *Xanthomonas* is less unstable than originally thought. Preservation of this strain in conventional agar slopes transfers did not show any deterioration of its xanthan production ability (9). The preservation of two strains (either a collection strain and an improved strain isolated in our laboratory) by serial transfer in slopes turned out to be a good method for culture maintenance. Along the 11 months tested, the strains showed a variability in xanthan production lower than 14% (with respect to the average and as measured in baffled shake flasks). In addition, the quality of the gum obtained practically did not vary along the time of study.

We have developed a novel method for the preservation of this bacterium in sterile cabbage seeds (8). The sterilization of *Brassica oleracea* seeds was achieved using gamma radiation and hypochloride washing while assuring that the seeds were viable (*i.e.* they were able to germinate). Bacterial viability showed oscillations but after 20 months it was 10 % of the initial. When these seeds were used as a pre-inoculum for a culture to produce xanthan, the final polymer concentration and the final apparent viscosity were, in the average of ten replicas, never lower than at the beginning of the experiment. The specific polymer production (per weight of final bacterial cells) increased about three-fold after 21 months of experimentation. The seeds method, besides being able to preserve the viability of the bacteria and their ability to produce xanthan in quantity and quality, has the advantages of an easy inoculation procedure, no need for transfers, less contamination risk and improved growth rate of the bacteria in the inoculation medium.

3.3 Broth rheology

The rheology of xanthan fermentation broths shows drastic changes during culture. Figure 4 shows typical rheograms (10). At high concentrations (*i.e.* above 15 kg m^{-3}) xanthan broths exhibit yield stress (11) and viscoelasticity (normal forces) (12). Although the yield stress concept has been controversial, it is a rheological property which plays an important role in mixing phenomena in xanthan fermentation. According to its definition, the existence of a true yield stress in xanthan solutions has been questioned. We have taken a practical approach and developed a technique for measuring the "apparent yield stress" which is useful for mixing purposes. The relaxation technique uses a low-cost Brookfield viscometer (11) and proved to be reliable and preferable to the fitting of rheological models (*i.e.* Casson, Herschel-Bulkley). Apparent yield stress values were used succesfully for the prediction of the size of the well mixed region in xanthan and Carbopol solutions in a stirred tank equipped with Rushton turbines of different diameters.

Viscoelastic properties have been extensively evaluated (12) in reconstituted broths as well as in various solutions of commercial xanthans. In general, we found that the higher the xanthan concentration, the higher the normal force at a given shear rate.

3.4 Mixing aspects

In collaboration with the University of Birmingham (U.K.), we have conducted several studies related to mixing and power consumption of a number of impeller configurations using simulated xanthan broths. Rushton turbines (13), the Lightnin A-315 (16) and the Scaba 6SRGT impeller (18) have been assessed in terms of the well mixed "caverns" and power drawn. A novel in-fermenter dynamometer (based in strain gauges and a telemetry system) has been used to assess power drawn in actual fermentations (14,15). Combinations of axial and radial impellers have been also assessed (17) using reconstituted xanthan broths having matched rheology to the actual ones. If compared with a single (D/T = 0.42) Rushton turbine, dual (D.T = 0.53) Rushton turbines allowed to increase 2.5 fold the efficiency of the process in terms of kg xanthan/kw-h.

Assessed in commercial xanthan solutions, the Lightnin A-315 impeller turned out to be suitable for mixing broths containing below 25 kg $m^{-3}$of xanthan (16). Interestingly, when this impeller was operated in the reverse mode (a kind of upwards pumping), better gas dispersion, less torque fluctuations, lower power drop due to aeration and larger cavern volumes was observed if compared at the same power drawn for the conventional (forward) downwards pumping mode. Drastic power drop and large torque instabilities were observed for highly concentrated xanthan solutions (35 kg m^{-3}).

The curved blades Scaba 6SRGT impeller is an interesting retrofitting possibility in xanthan fermentation. Our studies (18) have shown that the power consumption of a hollow blade agitator such as the Scaba 6SRGT is considerably less sensitive to aeration than a Rushton turbine in high viscosity simulated xanthan gum broths. This should allow more power to be spent on improving bulk mixing (*i.e.* larger well-mixed caverns) and mass transfer and should also improve the fermentation results. In addition, the lower power Scaba 6SRGT offers retrofitting with a larger D/T ratio whithout the need of changing the drive train of the agitation system as it would be the case with Rusthon turbines.

3.5 Heat treatment

After fermentation is completed, the *Xanthomonas* culture has to be inactivated. This is achieved by a heat treatment to the final broth which, besides to killing the bacteria, enhance the rheological properties of the xanthan in solution. Broths with low xanthan concentrations (< 10 kg m^{-3}) show important changes in rheology as a result of the heat treatment (10). Heated broths become more viscous and shear thinning (10, 15) and show higher apparent yield stress [Torres *et al* (20)] and different normal force patterns (12).

If compared with non-heated broths, the mixing of heated

broths results in considerably higher power drawn and more pronounced drop in power due to aeration (15).

3.6 Precipitation

Precipitation of xanthan gum from reconstituted fermentation broths has been studied (19) by our group. The influence of operation variables, such as broth's xanthan concentration, broth feeding rate and agitation speed, over the polymer yield, was established. Four impeller geometries were assessed and some of the results are shown in figure 5. Axial impellers exhibited decreased yields as stirring speed was higher whereas radial impellers showed the opposite though less pronounced trend. Clearly, mixing conditions play an important role in precipitation of xanthan gum as the yield and fiber size depend on it.

4. CONCLUSIONS

Xanthan gum production has been a biotechnological process which represented and excellent model of study requiring multidisciplinary approaches. The better understanding of biological and engineering aspects have been equally important in achieving process improvements.

5. ACKNOWLEDGEMENTS

Much of the work reported here has been supported by the National University of México (DGAPA-UNAM), the Mexican Science and Technology Council (CONACyT), the International Foundation for Science, the British Council, and the European Economic Community.

6. REFERENCES

1. Galindo, E., 1985, In: Prospectiva de la Biotecnología en México, Quintero, R. (Ed.) CONACyT-FJBS A.C., México D.F., pp. 65-92.
2. Sutherland, I.W., 1990, Biotechnology of Microbial Polysaccharides, Cambridge University Press, Cambridge, U.K.
3. Ramírez, M.E., Fucikovsky L., García-Jiménez F., Quintero, R. and Galindo, E., 1988, Appl. Microbiol. Biotechnol. 29, 5-10.
4. Torrestiana, B., Fucikovsky, L. and Galindo, E., 1990 Lett. Appl. Microbiol. 10, 81-83.
5. Torres, L.G., Brito, E., Galindo, E. and Choplin, L., 1993 J. Ferment. Bioeng. 75, 58-64.
6. Ramírez, M.E. 1993, M.Phil. Thesis, National University of México,Cuernavaca, México.
7. Galindo, E., Salcedo, G., Flores, C. and Ramírez, M.E., 1993, World J. Microbiol. Biotechnol. 9, 122-124.

8. Salcedo, G., Ramírez, M.E., Flores, C. and Galindo, E., 1992, Appl. Microbiol. Biotechnol. 37, 723-727.
9. Galindo, E., Salcedo, G. and Ramírez, M.E., 1994, Appl. Microbiol. Biotechnol. 40, 634-637.
10. Galindo, E., Torrestiana, B. and García-Rejón, A. 1989, Biopr. Eng. 4, 113-118.
11. Hannote, M., Flores, F., Torres, L. and Galindo, E., 1991, Chem. Eng. J. 45, B49-B56.
12. Torres, L.G., Nienow, A.W., Sánchez, A. and Galindo, E. 1993, Biopr. Eng. 9, 231-237.
13. Galindo, E., Nienow, A.W. and Badham, R. 1988, In: Proc. 2nd Bioreactor Fluid Dynamics Conf., King, R. (Ed.) BHRA The Fluid Engineering Centre, Cranfield, U.K., pp. 65-78.
14. Xueming, Z., Nienow, A.W., Kent, C.A., Chatwin, S. and Galindo, E., 1991, In: Procc. 7th European Conference on Mixing, Brugge, Belgium, vol. I, K.V.I. vzm, pp. 277-283.
15. Torrestiana, B., Galindo, E., Xueming, Z. and Nienow, A.W., 1991, Trans. I. Chem. E. 69(C), 149-155.
16. Galindo, E. and Nienow, A.W., 1992, Biotechnol. Prog. 8, 233-239.
17. Sánchez, A., Martínez, A., Torres, L. and Galindo, E., 1992, Process Biochem. 27, 351-365.
18. Galindo, E. and Nienow, A.W., 1993, Chem. Eng. Technol. 16, 102-108.
19. Albiter, V., Torres, L.G. and Galindo, E., 1994, Process Biochem. 29, 187-196.
20. Torres, L.G., Flores, F. and Galindo, E., 1994, Biopr. Eng. (in press).

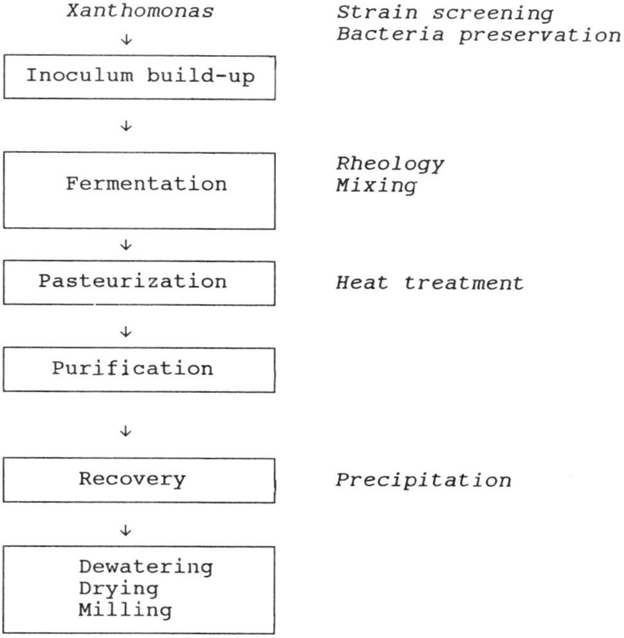

Fig. 1. Blocks diagram of the process for xanthan gum production indicating areas studied by the author's group and collaborators.

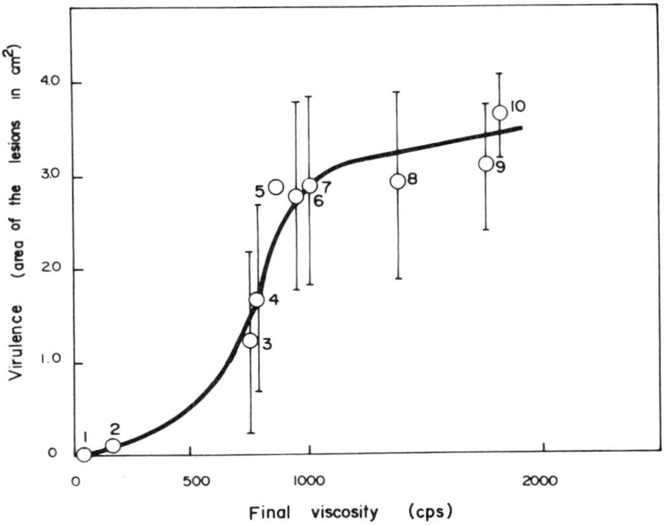

Fig. 2. Virulence as a function of final apparent viscosity of various *Xanthomonas* variants [after Ramírez *et al* (3)].

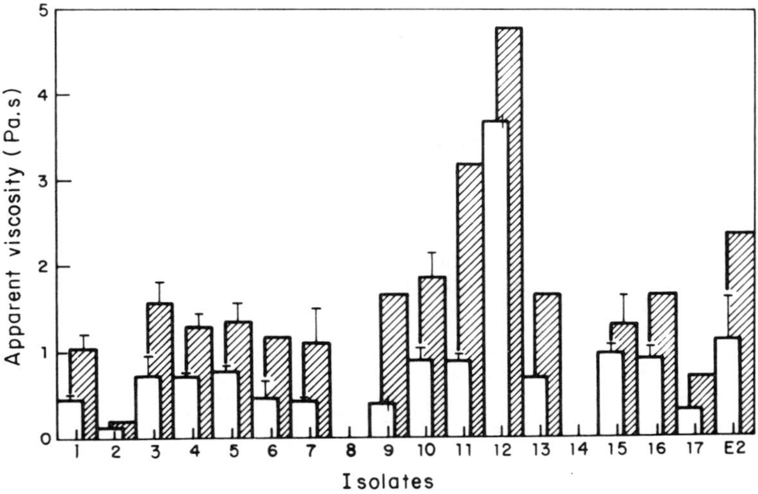

Fig. 3. Apparent viscosity of 17 different isolates of *X. campestris*. Open bars: viscosity before heat treatment. Closed bars: viscosity after heat treatment [after Ramírez (6)].

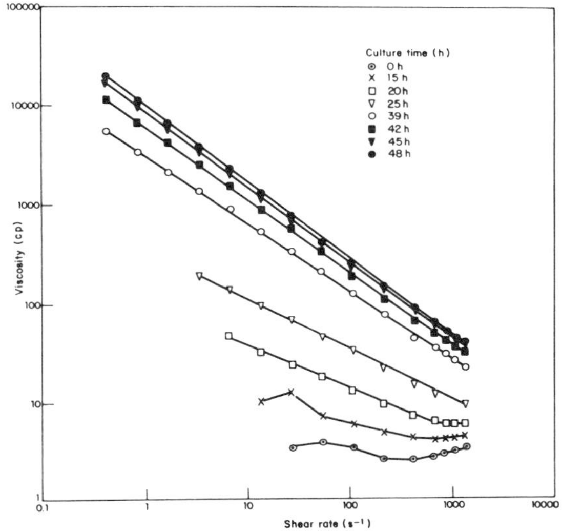

Fig. 4. Viscosity evolution during fermentation of *X. campestris* 1459-4L [after Galindo *et al* (10)].

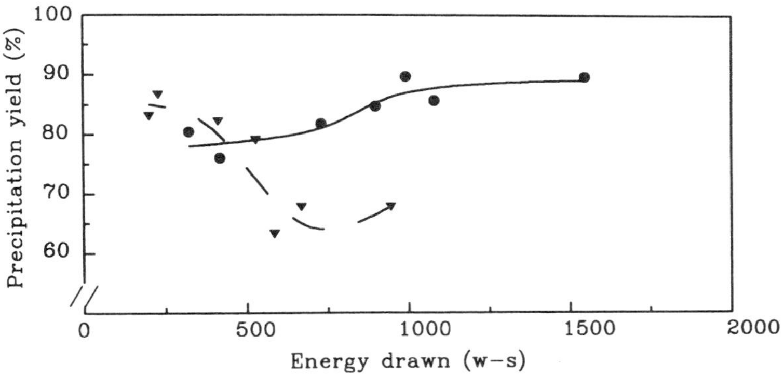

Fig. 5. Precipitation yield as a function of energy drawn and impeller type. (●) Rushton turbine (▽) marine propeller [after Albiter *et al* (19)].

A SHORT STUDY ON THE BIODEGRADATION OF WASTE WAX FROM A MARINE OIL TERMINAL

Suresh T. Nesaratnam
(Faculty of Technology, The Open University, Walton Hall, Milton Keynes, MK7 6AA, U.K.)

The "pigging" of crude oil pipelines carrying waxy crudes can result in significant quantities of waste wax, the disposal of which can be costly. In assessing the feasibility of reducing the disposal quantities, a short biodegradation study was undertaken at laboratory scale on a wax sample from an oil terminal. Polypropylene Pall rings coated with the wax were packed into two 4-litre glass fermenters for study. One fermenter was inoculated with Biolyte™ CX90 (a proprietary petroleum-degrading mixed culture) while the other acted as control. Both fermenters were aerated and wax degradation was monitored over 30 days. During this period, nearly all the short-chain alkanes (nC_9 - nC_{25}) in both fermenters were removed. The remaining components (long-chain alkanes and hexane-insoluble matter) were reduced by about a third. Overall, reduction in the weight of wax was 42.7% in the inoculated fermenter and 34.4% in the control. This study has application in the disposal of wax deposited in crude oil pipelines.

1. INTRODUCTION

Crude oil can often contain wax and this can deposit itself in the pipeline bringing the crude from a production site to the receiving terminal. Deposition of the wax results in decreased oil flows and hence is undesirable. So pipelines carrying waxy crudes are often cleaned with a bullet-shaped capsule (called a 'pig' in the industry). The 'pig' effectively scrapes off the wax and pushes it out at the end of the pipe in the oil terminal. In terminals handling waxy crudes, the amount of wax generated in this way can be massive. The wax from one particular terminal in the UK is disposed of by landfill - a costly operation when large, regular quantities are involved. Biotreatment at site to reduce the volume of the wax may be one option to adopt to reduce costs. With this in mind, a short biodegradation study was undertaken at laboratory scale on a sample of wax produced in the 'pigging' of an oil line bringing crude from the North Sea to the oil terminal concerned.

1.1 Literature Review

The microbial degradation of hydrocarbons has been recorded since the early part of this century Sohngen [1]but Zobell [2] was the first to show that nearly all the components of crude oil could be microbially degraded.

Zobell reported that there were more than 100 species of bacteria, yeasts and fungi capable of oxidising hydrocarbons. Generally, each species attacks a certain type of hydrocarbon; thus a mixed culture would probably be best to degrade a complex hydrocarbon mixture such as crude oil. Miget [3] found that mixed cultures of hydrocarbon-degrading bacteria were more effective in biodegrading oil than individual isolates. Hughes and McKenzie [4] found that amongst hydrocarbon-degrading cultures, the pseudomonads, *Achromobacter, Flavobacter* and *Corynebacter* dominated the bacteria; *Norcardia,* mycobacteria and streptomyces are also frequently found. A range of yeasts, *Candida* species especially, and moulds such as *Penicillium* and *Cladosporium* are also quite common.

Hydrocarbon-degrading microbes have been found in a diversity of environments, e.g. in garden soil, compost, estuaries, oceans, marine sediments and lakes Brown [5]. The largest numbers have been found in environments associated with oil Ahearn *et al* [6].

The short-chain alkanes ($<C_{10}$) are the first compounds attacked by micro-organisms Brown [5]. The short-chain branched alkanes are the next to be metabolized; the greater the chain length and the amount of branching, the more resistant the compound is to microbial attack. After these

components, the aromatic compounds are oxidized, in the order monocyclic, bicyclic, tricyclic and polycyclic Wyman and Brown [1].

2. METHODOLOGY

A sample of the wax waste was analysed in the laboratory. The results are shown in Table 1.

Table 1: Composition of the waste wax

Component	Concentration (%)	Component	Concentration ($\mu g\ g^{-1}$)
C	~85	Cd	<0.44
H	~14	Cr	<2.0
N	<0.3	Cu	4.1
P	<0.3	Fe	150
S	~0.6	Pb	<4.4
Water	<3	Ni	2.2
Ash	0.4	Va	8.3
Asphaltene	0.22	Zn	0.6

Gas chromatographic analysis of the wax revealed that it was made up largely of n-alkanes with only trace quantities of aromatics and branched alkanes.

For the biodegradation trial, high specific surface area polypropylene Pall rings (1.5 cm diameter) were dipped into molten wax, dried and then packed into two glass 4-litre fermenters of the type shown in Fig. 1. One litre of tap water was added to each fermenter. There was 143 g of wax in Fermenter 1 and 168 g in Fermenter 2, contained on 750 Pall rings.

The first fermenter was inoculated with 33 g of Biolyte™ CX 90, a proprietary mixed culture which was purported to initiate and accelerate the biodegradation of petroleum hydrocarbons, obtained from International Biochemicals (UK) Ltd., of Slough, Berkshire, England.. The mixed culture consisted of cells of *Bacillus*, *Pseudomonas*, *Actinomyces* and yeasts, together with a small quantity of surfactant, all on a cereal base.

33 g of the same mixed culture additive was sterilized and added to the second (control) fermenter.

The C:N:P ratio in the fermenters was adjusted to 100 : 8.2 : 1.2 using $NH_4\,NO_3$ and $NH_4\,H_2\,PO_4$ dissolved in 2 litres of water. The media temperature in the fermenters was maintained at about 25 °C, and each fermenter was aerated by two aquarium air pumps delivering a total of 10 litres of air a minute.

The exhaust air from the fermenter was bubbled through a solution of germicide before discharge to atmosphere.

On days 10, 20 and 30 ten Pall rings were taken for analysis from the middle of each fermenter. Half the culture broth was also withdrawn and replaced with tap water containing N and P as nutrient. Cell counts and temperature and pH readings were taken. The wax from five of the rings was analysed by gas chromatography.

3. RESULTS

The cell counts in the fermenters did not differ significantly (Fig. 2), implying that there were resident microorganisms in the wax itself giving high values in Fermenter 2.

Gas chromatographic analysis of the wax revealed an homologous series of n-alkanes ranging in carbon number from nC_9 - nC_{43} (Fig. 3). The standard used in the determination was squalane ($nC_{30,}$ marked "S" in Fig. 3)The n-alkanes formed two distinct envelopes with maxima at nC_{14} and at nC_{35}. A series of isoprenoidal alkanes was also present. There was, in addition, a hexane-insoluble fraction in the wax. This was thought to be a very high molecular weight substance.

Each set of 5 rings analysed on Days 10, 20 and 30 was considered for its content of short-chain alkanes (nC_9 - nC_{25}), long-chain alkanes (nC_{26} - nC_{43})and hexane-insolubles. The data are shown in Table 2 and Fig. 4.

Table 2: Results from the Two Fermenters (weights represent quantities on 5 Pall rings)

		FERMENTER 1			FERMENTER 2		
	Day 0	Day 10	Day 20	Day 30	Day 10	Day 20	Day 30
Short-chain alkanes (nC_9 - nC_{25})	40.64 mg	20.39 mg	6.03 mg	1.82 mg	23.36 mg	7.05 mg	3.04 mg
Percentage of Original Remaining	100	50.2	14.8	4.5	57.5	17.3	7.5
Long-chain alkanes (nC_{26} - nC_{43})	617.42 mg	619.47 mg	550.01 mg	375.59 mg	542.56 mg	591.22 mg	428.85 mg
Percentage of Original Remaining	100	100.3	89.1	60.8	87.9	95.8	69.5
Total Wax	658.06 mg	639.86 mg	556.04 mg	377.21 mg	565.92 mg	598.27 mg	431.89 mg
Percentage of Original Remaining	100	97.2	84.5	57.3	86.0	90.9	65.6

3.1 Removal of short-chain alkanes (nC_9 - nC_{25})

The short-chain alkanes were easily degraded (Fig. 4) being reduced to 4.5% and 7.5% of their original content after 30 days in Fermenters 1 and 2, respectively. Their removal in the control (uninoculated) fermenter indicates that the native microorganisms on the wax were capable of easily biodegrading these compounds. Fermenter 1, which had the proprietary mixed additive, performed marginally better. The short-chain alkanes were only a small proportion (6.18%) of the wax. Figure 5 shows the gas chromatograph of the wax after 30 days in Fermenter 1. The loss of the short-chain alkanes is evident. The GC trace for Fermenter 2 was similar.

3.2 Removal of long-chain alkanes (nC_{26} - nC_{43})

There was a lag phase before the removal of these compounds was initiated. It is conceivable that the microorganisms were adapting their enzyme systems to these complex compounds during this period. As expected, the degree of removal was not as high as for the short-chain alkanes. After 30 days, there were 60.8% and 69.5% of the long-chain alkanes and hexane-insolubles remaining in Fermenters 1 and 2, respectively.

3.3 Overall reduction in mass of wax.

After 30 days, 42.7% of the wax in Fermenter 1 was found to have been removed, while in Fermenter 2 removal was 34.4.%. While it does seem that the addition of a proprietary culture designed for the particular waste is beneficial to its degradation, the above results show that the native microorganisms are not to be disregarded. Further studies involving optimization of fermenter conditions coupled with careful economic analysis will indicate the option to use in a full-scale system.

3.4 Isoprenoidal alkanes and biodegradation.

Amongst the isoprenoidal alkanes of the wax were pristane (C_{19}) and phytane (C_{20}). These compounds are more resistant to microbial degradation than the normal alkanes and ratios of pristane and phytane to their associated C_{17} and C_{18} normal alkanes (heptadecane and octadecane) may be used as indicators of the degree of degradation the n-alkanes have undergone. These ratios are shown in Table 3.

As can be seen from the table, the nC_{17}/Pr and nC_{18}/Ph ratios decreased till day 20, indicating that degradation was taking place. The ratios increased on day 30 but this was due to the fact that the pristane and phytane contents themselves were dramatically reduced, most likely by microbial action.
Comparing the ratios of nC_{17}/Pr and nC_{18}/Ph on days 10 and 20 for the two fermenters indicates that the rate of degradation of the C_{17} and C_{18} compounds was higher in Fermenter 1 in the first 10 days. Thereafter the microorganisms in Fermenter 2 appear to have adapted to the wax with the result that on day 20 ratios for both fermenters were similar.

Table 3: Ratios of Heptadecane to Pristane (nC_{17}/P_r), Octadane to Phytane (nC_{18}/Ph), and Pristane to Phytane (Pr/Ph) in wax from the Pall rings.

		FERMENTER 1			FERMENTER 2		
	Day 0	Day 10	Day 20	Day 30	Day 10	Day 20	Day 30
nC_{17}/Pr	2.02	0.61	0.11	0.25	1.0	0.17	0.24
Weight of Pr (mg)	1.43	1.61	1.03	0.32	1.54	1.11	0.46
nC_{18}/Ph	1.72	0.58	0.11	0.14	0.82	0.14	0.21
Weight of Ph (mg)	1.45	1.75	1.16	0.35	1.66	1.17	0.52
Pr/Ph	1.17	0.92	0.89	0.93	0.93	0.95	0.98

4. CONCLUSIONS

In a 30-day laboratory scale biodegradation trial with waste wax from a marine oil terminal, it was found that decomposition of the wax was possible. In a fermenter containing a proprietary mixed culture of microorganisms adapted for the degradation of petroleum hydrocarbons, the mass of wax was reduced by 42.7%, while in the control fermenter the percentage reduction was 34.4%. The short-chain alkanes (nC_9 - nC_{25}) in the wax were almost totally eliminated in both the fermenters. The other major components of the wax were the long-chain alkanes (nC_{26} - nC_{43}) and a fraction which was hexane-insoluble. These components were reduced by about a third.

5. ACKNOWLEDGEMENT

The Institute of Offshore Engineering at Heriot-Watt University in Edinburgh, Scotland, is gratefully acknowledged for allowing the data in this paper to be presented.
Acknowledgement is due also to International Biochemicals (UK) Ltd., for providing samples of the mixed culture, Biolyte™ CX90 used in the above study.

6. REFERENCES.

1. Sohngen, N. L., (1905). Zentr Bakt Part 15, II Abt, p513. Chem Zentr I, p949.1.
2. Zobell, C. E., (1969). Microbial modification of crude oil in the sea. In Proceedings of the AP/FWPCA Joint Conference on Prevention and Control of Oil spills. American Petroleum Institute, Washington, DC.
3. Miget, R. J., (1972). Bacterial seeding to enhance biodegradation of oil slicks. In Proceedings of the Workshop on the Microbial Degradation of Oil Pollutants. Georgia State University, Atlanta.
4. Hughes ,D. E., & McKenzie P., (1975). The microbial degradation of oil in the sea. Proc. R Soc., London, B 189, pp375-390.
5. Brown, L. R., (1987). Oil degrading micro-organisms. Chem. Eng Prog. Oct, pp35-40.
6. Ahearn, D. G., Meyers S. P., & Standard P. G,. (1971). Dev. Ind Microbial 12,p 126.
7. Wyman, J. R., Brown, L. R., (1975). Dev Ind Microbial 17, p311.

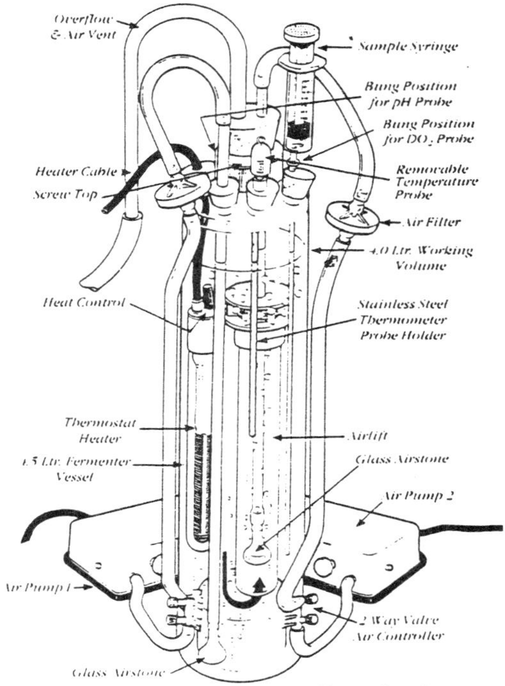

(Reproduced courtesy of Dr. R.N. Greenshields of GB Biotechnology Ltd., Wales).

Fig. 1 : The type of fermenter used in the study : the fermenter was packed with Pall rings for the study.

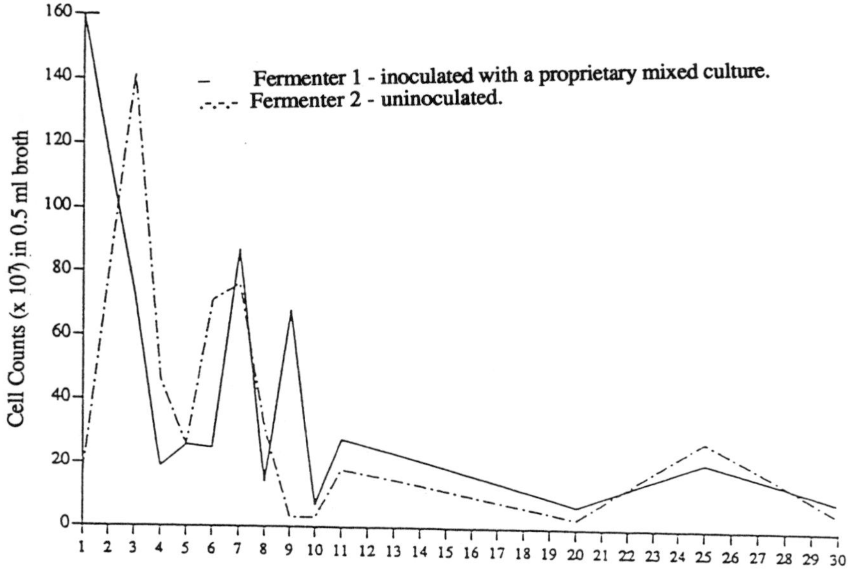

Fig. 2 : Cell counts in the fermenters.

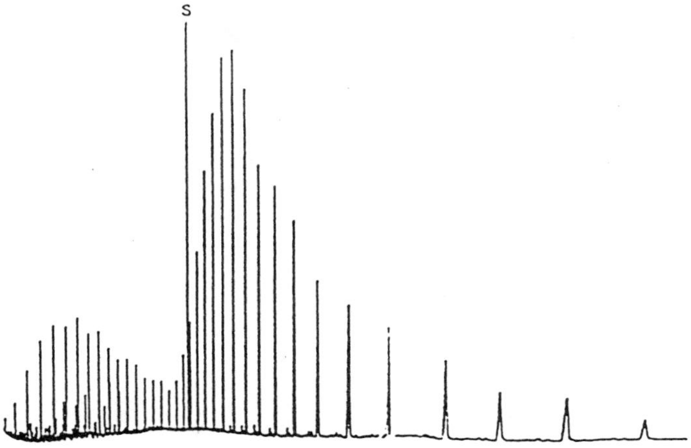

Fig. 3 : GC trace of wax at start of experiment showing an homologous series of n-alkanes with squalane (nC_{30}) as standard

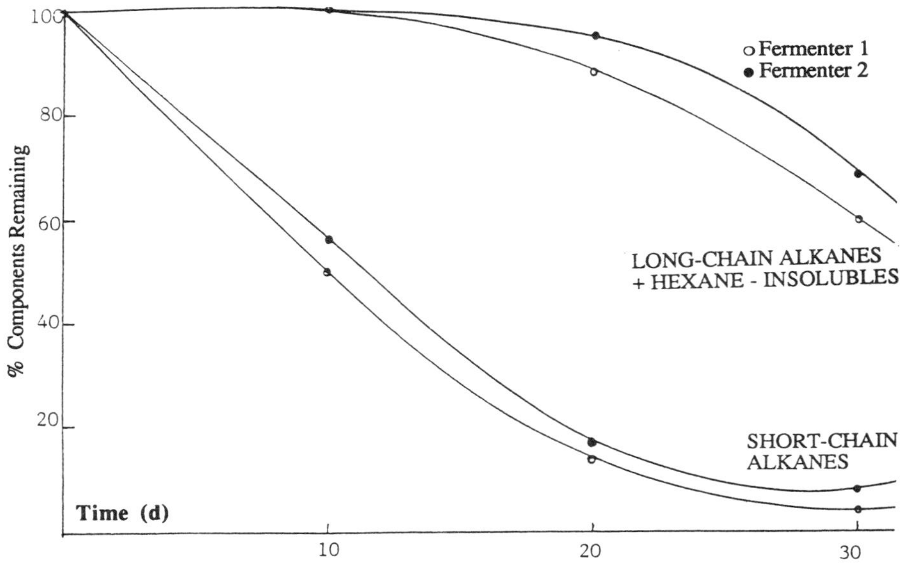

Fig. 4 : The degradation of the wax

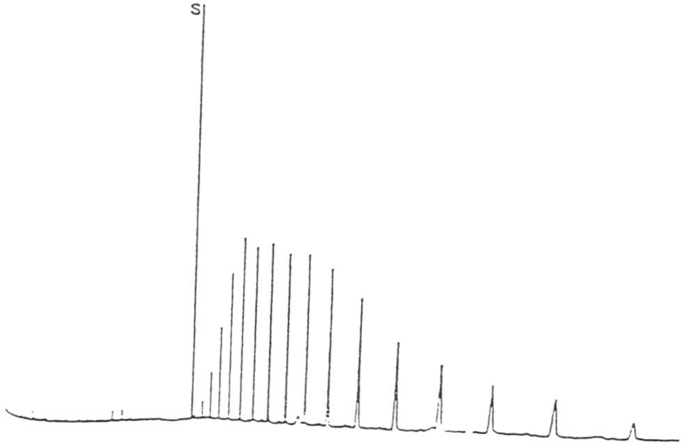

Fig. 5 : GC trace of wax from Fermenter 1 after 30 days.

MODEL PREDICTION AND VERIFICATION OF A TWO-STAGE HIGH-RATE ANAEROBIC WASTEWATER TREATMENT SYSTEM SUBJECTED TO SHOCK LOADS

M. Romli, J. Keller, P.L. Lee and P.F. Greenfield

Department of Chemical Engineering, The University of Queensland, St. Lucia, Qld 4072, Australia

The dynamic responses of an anaerobic wastewater treatment system subjected to shock loads were determined experimentally and simulated using a mechanistic model. The comparison between the measured and the predicted data showed the ability of the model to predict the dynamic behaviour of most reactor variables. Of particular importance was the prediction of the organic acids accumulation in the acidification reactor. It was shown that lactic acid was an important intermediate fermentation product in anaerobic degradation processes handling carbohydrate-based wastewater.

1. INTRODUCTION

The development of a dynamic model which is reliable to describe the overall mechanisms of anaerobic wastewater treatment processes can provide a useful tool for design, control, optimization, and operational evaluation. Through model simulation, different operational strategies can be evaluated without the need to involve actual plant operations, which therefore saves time and money. A mechanistic model of anaerobic degradation processes handling carbohydrate-based wastewater has been recently developed (Costello *et al.*, 1). The model has been verified against the experimental data of single-stage anaerobic configurations. Further modifications and improvements of the model have allowed the model to be used for the simulation of two-stage anaerobic wastewater treatment. This improved model has been shown to predict the two-stage reactor behaviour with and without recycle operation and to describe the effect of pH manipulation of the acidification reactor (Romli, 2). In this present study, the ability of the model to describe the dynamic response of a two-stage high-rate anaerobic wastewater treatment system subjected to a step increase in feed concentration and flow rate will be discussed.

2. MATERIALS AND METHODS

2.1 Experiment

The experiments were carried out in a laboratory scale two-stage high-rate anaerobic wastewater treatment system, consisting of a continuously stirred tank reactor

(CSTR) as the acidification reactor and a fluidized sand bed reactor (FBR) as the methanogenic reactor (Figure 1). The liquid volume of the acidification reactor was 1.64 litre and that of the methanogenic reactor was 3.30 litre. Throughout the experiments the acidification reactor was controlled at a pH of 6.0 by automatic addition of sodium hydroxide, whereas the methanogenic reactor was only monitored. Both reactors were operated at a constant temperature of 35°C. No recycle was employed. A synthetic wastewater consisting of dilute sugar cane syrup enriched with nitrogen, phosphorous and trace metal elements was used as an influent.

The shock load experiments were conducted by introducing a 12 hour 100% step increase in feed concentration. After the normal feed was resumed and the system had reached steady state conditions, an 8 hour 100% step increase in feed flow rate was then introduced. The normal volumetric loading rate applied to system was 22.6 kg COD/m^3.day. For a total time of at least 24 hours, reactor variables such as pH of the second reactor, alkali flow rate, gas generation rate, gas composition and effluent organic acids concentration were measured at hourly intervals. The details of reactor configuration and analytical methods for these variables were described elsewhere (Romli, 2).

2.2 Simulation

An improved version of a mechanistic model of anaerobic degradation processes developed initially by Costello *et al.* (1) was used to simulate the reactor performance during the shock load situations. This model was implemented in a recently developed simulation package NIMBUS (Newell and Cameron, 3). In NIMBUS the model was set up in a modular form according to the treatment plant flowsheet (Figure 2). To use this dynamic model three types of model parameters, namely experimental, physico-chemical and biological parameters need to be defined. The experimental parameters were given according to the experimental operating conditions. The physico-chemical parameters were based on the values reported by Costello *et al.* (4). The biological model parameters were based on the values reported by Costello *et al.* (4) and Pavlostathis and Giraldo-Gomez (5). The differential and algebraic model equations were solved simultaneously using a Sparse Matrix Version of the Diagonally Implicit Runge-Kutta (DIRKS) solvers (Newell and Cameron, 3), with an integration accuracy of 0.0001. Further details about the model structure and parameters used can be found elsewhere (Romli, 2).

3. RESULTS AND DISCUSSION

The dynamic responses of some of the reactor variables to the concentration and hydraulic shock loads are presented in Figures 3-8. Both the measured (represented as symbols) and the predicted (represented as solid lines) results are displayed. Error bars shown in some figures were the measurement uncertainties at a 95% confidence level. The vertical dotted line in each figure indicated the duration of shock load.

Figure 3 shows the dynamic behaviour of alkali flow rate in response to the step increase in feed concentration. As expected, the alkali required to maintain the pH setpoint of the acidification reactor increased during the course of the shock load from 40 to 90 ml/h. The maximum value was reached in less than 2 hours after the disturbance was introduced. A similar response was shown in the hydraulic shock load, in which the alkali flow rate was doubled from 40 to 80 ml/h (Figure 4). In both situations, a rapid recovery (in less than 4 hours) was observed as soon as the normal feed conditions were restored. The model was shown to predict this variable quite well. The discrepancy in the maximum value of alkali flow rate during the course of the shock loads was most likely due to the difference in the performance of pH controller used in the experiment and in the simulation. The concentration shock load, as shown in Figure 3, decreased the effluent pH of the methanogenic reactor from 7.5 to 7.3. It was also noted that the pH started to recover before the shock load was terminated. On the contrary, the hydraulic shock load led to a more severe pH drop and also slower pH recovery. These effects were followed closely by the model prediction.

The effect of both concentration and hydraulic shock loads on the gas generation rate of the acidification reactor was only marginal. The gas generation rate in the methanogenic reactor increased considerably during the shock loads (Figure 5). The model was shown to predict this variable reasonably well. The unusual response at the end of the shock load was not typical; it could possibly be as a result of the effervescent of bubbles during the periods of high gas production which then enhanced the gas transfer from the liquid phase. The discrepancy between the measured and the predicted data for the gas phase variables in the first reactor was due to the low gas production rate and the large head space in this reactor. The concentration shock load increased the concentration of carbon dioxide in the biogas in both reactors. A similar response was also observed during the hydraulic shock load as shown in Figure 6. Again, the predicted data showed reasonable agreement with the experimental results.

The effects of a step increase in feed concentration and flow rate on the effluent organic acids concentration of the acidification reactor are presented in Figures 7 and 8, respectively. During steady state conditions, the predominant organic acids found in the effluent of the first reactor were acetic, propionic and butyric acids. Other organic acids such as lactic and formic acids were detected at low levels. It should be pointed out here that the biological model parameters used in this simulation were fitted initially for molasses feed, in which ethanol was detected at insignificant levels (lower than butyric acid). It was found during the present study that ethanol was detected at levels 2.5 times higher. This fact might explain the overprediction of propionic acid shown in both figures, because this fermentation product was not included in the model. The significant effect of the concentration and hydraulic shock loads was the considerable accumulation of lactic acid. This species was modelled quite accurately as demonstrated in both figures. This result thus supported the inclusion of lactic acid as an important intermediate product in anaerobic processes handling carbohydrate-based wastewater. It also verified the use of hydrogen inhibition and regulation to describe the preference of the microbial fermentation pathway under the limitation of electron carriers, in particular the oxidized form of nicotinamide adenine dinucleotide (NAD). The formation of lactic acid is the simplest solution for finding an electron acceptor, because pyruvic acid produced during the EMP pathway acts as the acceptor. As shown in Figure 7, acetic, propionic and butyric acids increased steadily in response to the concentration shock load. A further increase observed after the feed shock load terminated, especially for acetic and propionic acids, was a result of the lactic acid degradation. This dynamic behaviour was also predicted by the model. In contrast to the response shown in the concentration shock load, the hydraulic shock load led to a decrease in acetic acid. This was due to a combination of the dilution from the influent and also the fermentation shift towards the production of lactic acid. The model prediction was in agreement with this experimental result. The effect on butyric acid species as shown by the model prediction and the experiment was only marginal.

4. CONCLUSION

The applicability of the improved version of a mechanistic model to describe the reactor performance of a two-stage high-rate anaerobic wastewater treatment system subjected to a short-term of concentration and hydraulic shock load situations was demonstrated. The dynamic prediction of alkali flow rate, final effluent pH, gas generation rate and composition as well as the effluent organic acids concentration was in reasonable agreement with experimental data. The model therefore provides a useful tool for design, optimization, and operational evaluation purposes of two-stage anaerobic reactor configurations.

5. ACKNOWLEDGMENTS

The financial support provided by the IUC/WB XVII (to M.R.) and also by the Swiss National Science Foundation (to J.K.) is gratefully acknowledged.

6. REFERENCES

1. Costello, D.J., Greenfield, P.F., and Lee, P.L., 1991, Wat. Res., 25, 847-858.
2. Romli, M., 1993, PhD dissertation, The University of Queensland, Australia.
3. Newell, R.B. and Cameron, I.T., 1991, NIMBUS User's Manual. CAPE Centre, The University of Queensland, Australia. pp. 133-144.
4. Costello, D.J., Greenfield, P.F., and Lee, P.L., 1991, Wat. Res., 25, 859-871.
5. Pavlostathis, S.G. and Giraldo-Gomez, E., 1991, Crit. Rev. Env. Cont., 21(5,6), 411-490.

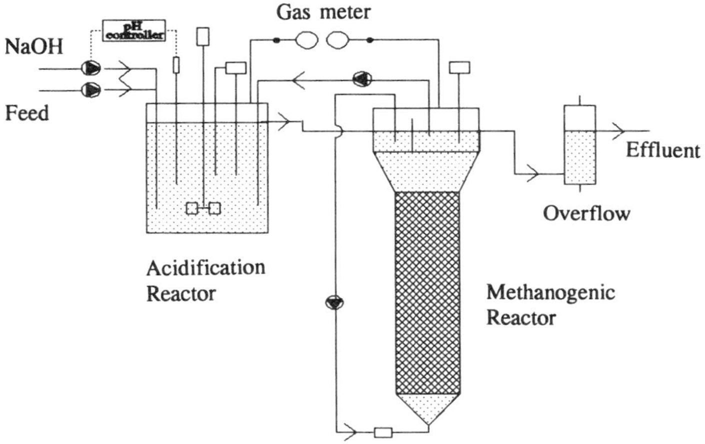

Figure 1. Schematic diagram of a two-stage high-rate anaerobic wastewater treatment system

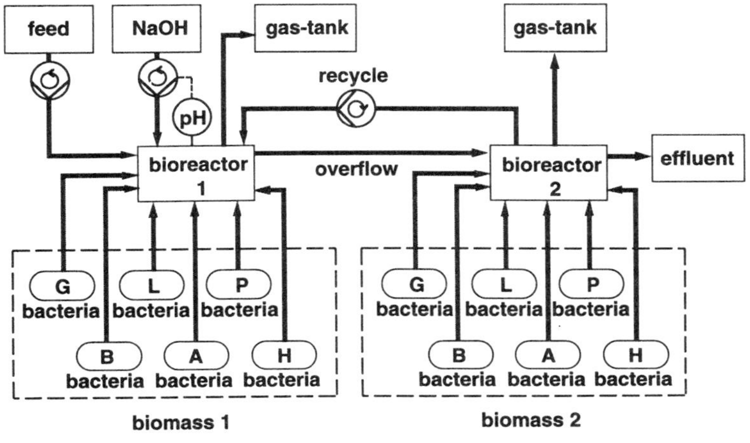

Figure 2. Model configuration of a two-stage anaerobic wastewater treatment system in NIMBUS

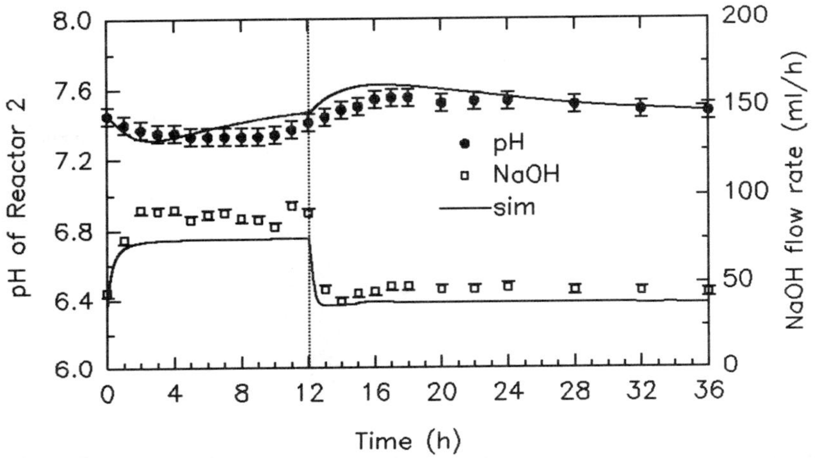

Figure 3. Effect of a concentration shock load on alkali flow rate and pH of Reactor 2

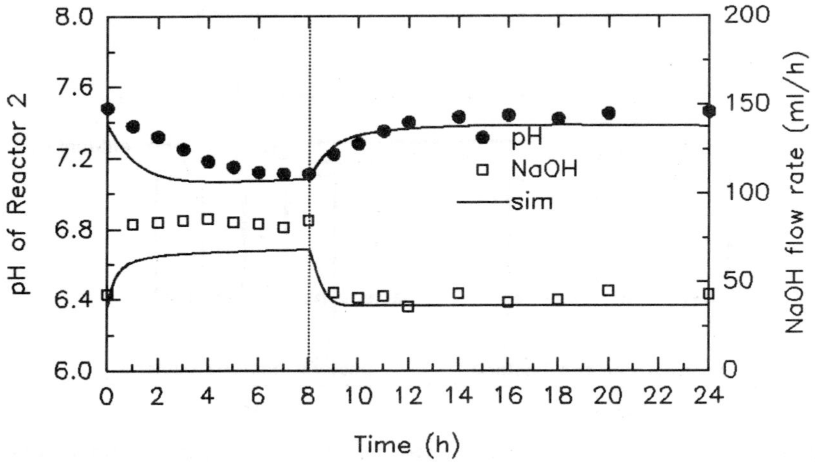

Figure 4. Effect of a hydraulic shock load on alkali flow rate and pH of Reactor 2

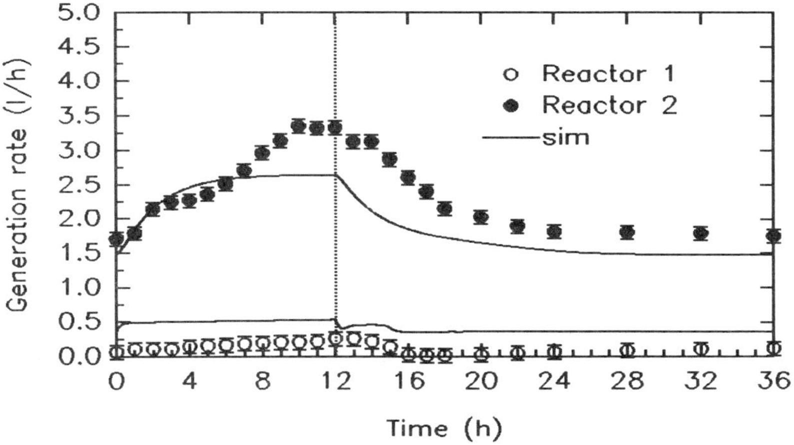

Figure 5. Effect of a concentration shock load on gas generation rate

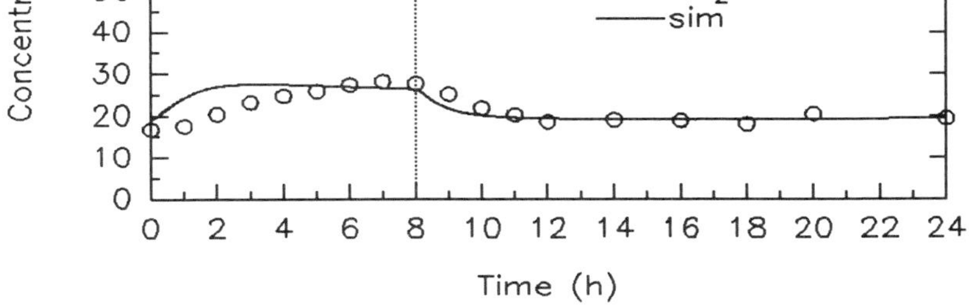

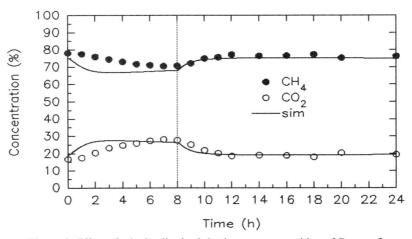

Figure 6. Effect of a hydraulic shock load on gas composition of Reactor 2

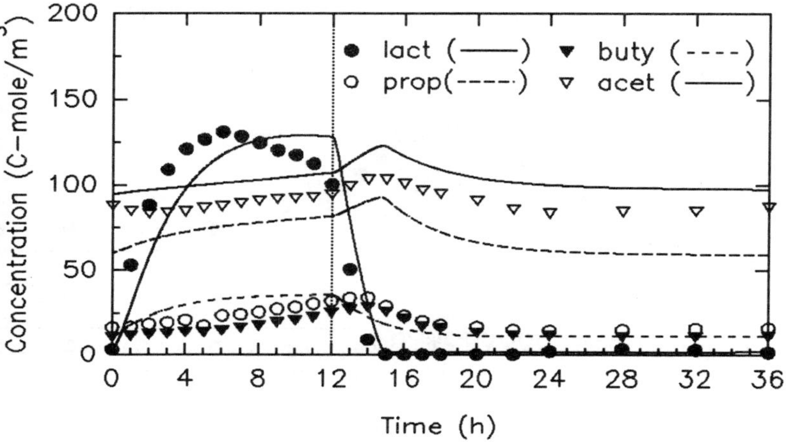

Figure 7. Effect of a concentration shock load on effluent organic acids of Reactor 1

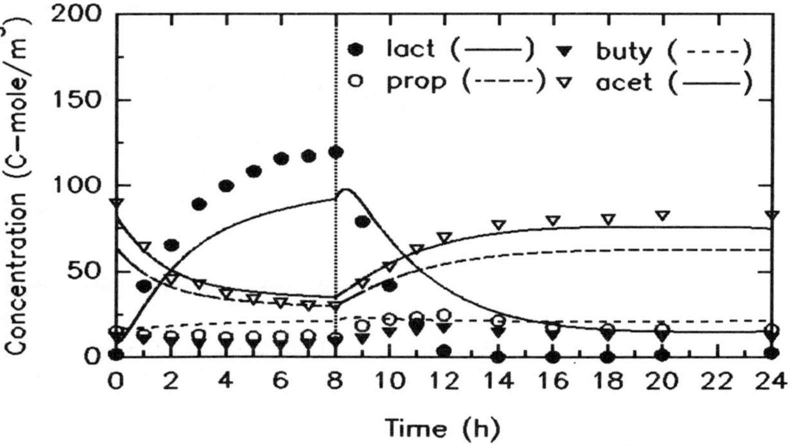

Figure 8. Effect of a hydraulic shock load on effluent organic acids of Reactor 1

THE DEGRADATION OF OIL PALM TRUNK USING ENZYMES FROM TRICHODERMA REESEI QM 9414.

Putri Faridatul Akmar, Azizol Abdul Kadir and K.C Khoo
Forest Research Institute Malaysia, Kepong, 52109 Kuala Lumpur

The suitability of using oil palm trunk as substrate for enzyme production and saccharification was studied. The process was carried out continously using a two-packed-reactor in series. The ß-glucosidase activity in the enzyme production reactor reached a maximum of 0.00034U/ml on the seventh day whilst the FPase reached maximum of 0.018U/ml on the ninth day and CMCase at 0.12U/ml on the eighth day. In the saccharification reactor, the enzymes produced successfully converted the oil palm trunk substrate into a maximum of 500μg/ml of glucose. This study showed that the oil palm trunk is suitable to be used as substrate for enzyme production and saccharification simultaneously.

1. INTRODUCTION

There is an abundance of oil palm trunk generated annually during replanting when the oil palm tree reaches the age of 25 years and above and its fruits are no longer economical to harvest. The oil palm trunk has not been fully utilized despite several attempts to convert it into value-added products. With the stringent zero-burning regulations and the abundance of trunks, more ways of utilization are needed urgently.

The oil palm trunk rich in sugars and carbohydrates, is easily attacked by fungi and insect pests (1). However, it is not advisable to leave the trunks on the plantation site to degrade by itself as it encourages the proliferation of rhinocerous beetles which will eventually attack the newly planted palm. Furthermore the rate of degradation is very slow and it will take more than a year for the unprocessed trunk to be fully decomposed naturally (2).

This study looks into several aspects of degradation of oil palm trunk using enzymes from the fungus Trichoderma reesei, grown on the oil palm trunk substrate, and the conversion of the substrate into fermentable sugars. The study was carried out using two column packed-reactors which were set up in series.

2. MATERIALS AND METHOD

2.1 Substrate

The oil palm section about 9.5m from the ground level and 40% radius was used as the substrate for this study. The disk was chipped, milled and sieved through a 200mesh size.

2.2 Inoculum

The Trichoderma reseei QM9414 was obtained from the UKM culture collection. It was grown on patato dextrose agar (PDA) for 6 days in 30°C incubator. Spores were collected using sterile 0.1% v/v a Tween 80. Only 200ml of the spores were used as inoculum.

2.3 Medium

Mandel's medium (3) was used. The composition of the medium is as shown in Table 1.

Table 1: The composition of medium used for the growth of fungus and secretion of enzymes.

chemical	g/L
$(NH_4)_2SO_4$	1.4
$MgSO_4.7H_2O$	0.3
KH_2PO_4	2.0
$CaCl_2.2H_2O$	0.4
D-Glucose	0.05
$FeSO_4.7H_2O$	0.005
$MnSO_4.H_2O$	0.0016
$ZnSO_4.7H_2O$	0.0014
$CoCl_2$	0.002
EDTA $Na_2.2H_2O$	0.05

2.4 Process

The process was carried out in two glass-column reactors, with a working volume of 1600ml and packed with stainless steel sponge, set up in series. About 50g of ground oil palm trunk substrate was arranged in step-wise formation in both the reactors (Figure 1) to allow maximum contact and smooth flow of medium.

The inoculum was passed through the reactor to allow spores to be trapped on the substrate and within reactor voids. Initially the first reactor was filled with medium containing glucose to allow the fungus to grow for 48hr after which medium without glucose was pumped in at the flow rate of 0.1ml/min. A continous flow-through operation was then started with an input medium and effluent flow rate maintained at 0.1ml/min in the second reactor which contained the substrate in acetate buffer of pH 4.8.

The first reactor was operated at 30°C, suitable for the growth of fungus and secretion of enzyme while the second reactor at 50°C was used for the saccharification process. Air was supplied into both reactors at the flow rate of 1600ml/min. The process was carried out over 26 days and samples of the effluent stream from each reactor were taken every 8 hrs.

2.5 Analysis

Samples from the effluent stream of each reactor were analysed for dissolved protein content (4), reducing sugar (5), filter paper activity (6), ß-glucosidase activity (7) and carboxymethyl cellulase activity (8). The pH was also recorded.

3. RESULTS AND DISCUSSION

This type of reactor set-up where the saccharification process is separated from the growth process, enhances the growth of the fungus for a higher secretions of enzymes. The initial 48hr batch process is ideal for the fungus to grow on the glucose medium. When the glucose content in the reactor reached its minimum, the fungus will feed on the cellulose of the substrate and secrete enzymes at the same time for their maintainance. Mitra and Wilke (9) found out that increase in growth of mass of the fungus would double up the enzyme productivity provided other conditions including acidity are suitable. At high acidity poor enzyme activities will result as fungus become autolyzed. Normally there is a lag phase about 30 hrs between fungal growth and enzyme secretion.

During the growth of the fungus the pH in the first reactor fell to 2.8 two days after inoculation from an initial of 5.5 and fell further to 1.4 on the 9th day (Figure 2). The pH, however, increased to 2.4 on the 11th. day and increased further to 5.7 after the 21st. day. The protein content too increased and reached a maximum of 820.5 μg/ml on the third day. According to Mandels (10), the fungus Trichoderma normally releases some acid as enzymes are secreted. The situation here, however, could not be used as the indication of the growth of fungus or secretion of enzyme because the changes in pH and protein were mainly due to the high soluble protein and acid content of the oil palm trunk solubilized matter (11). These intefered with the conditions in the reactor and resulted in low enzyme activities (Figure 3).

The ß-glucosidase activity was detected earlier on the 7th. day and reached a maximum 0.00034U/ml whilst the Fpase reached a maximum on the 9th day and the CMCase at 0.12U/ml on the 8th day. In the second reactor the enzyme activities were too low to be detected.

Even though the activities of the enzymes produced were low, the enzymes were able to degrade the carbohydrates of the oil palm trunk substrate in the second reactor. As shown Figure 2 the glucose content in the second reactor increased rapidly from 250μg/ml up to 500μg/ml on the 9th day and decreased steadily thereafter. The changes in the pH profile of the second reactor were due to the acid in the first reactor being fed into the second reactor. The protein was high as it was being accumulated from the first reactor.

The degradation of the substrate was ascertained by the use of SEM and as shown from the micrograph obtained (Figures 4 & 5). In Figure 4, most parenchymatous tissues and vascular bundles for the substrate from the second reactor were already attacked or broken down as compared to Figure 5 of the substrate from the first reactor where the parenchymatous cells could still be seen after the process. The data of the weight loss (Table 2) too supported this evidence.

Table 2: Weight loss of substrate during both processes (percent of the dry matter of the starting material).

substrate	% weight loss
of reactor 1	49.26
of reactor 2	60.26

4. CONCLUSION

This study demonstrated that the oil palm trunk was easily degraded by the cellulase of Trichoderma reesei. The degradation could yield more fermentable sugars if the activities of enzymes were higher. The process in the first reactor could be optimized for greater enzyme activities by removing the soluble protein and acids of the oil palm trunk.

5. REFERENCES

1. Tomimura, Y., 1992 JARQ 25:283-288.

2. Putri Faridatul Akmar, 1993 FRIM Terminal report for project 181.

3. Theodorou, M.K., M.J. Bazin, & A.P.J. Trinci, 1981 Arch. Microbiol. 130:370-380.

4. Lowry, O.H., N.J. Rosebrough, A.L. Farr & R.J. Randall, 1951 J. Biol. Chem. 193:265-275.

5. Somogyi, M., 1952 J. Biol Chem. 195:19-22.

6. Mandels, M, R. Andreotti & C. Roche, 1976 Biotechnol Bioeng Symp 6:21-33.

7. Theodorou, M.K., M.J. Bazin, & A.P.J. Trinci, 1980 Trans Bri Mycol Soc 75:45-54.

8. Wood, T.M. & McCrae, 1977 J. Biochem 128:1183-1192.

9. Mitra G. & Wilke R., 1975 Biotechnol Bioeng 17:1-13.

10. Mandels, M., 1971 Appl Microbiol:152-154.

11. Putri Faridatul Akmar, 1993. MSc thesis. Universiti Kebangsaan Malaysia.

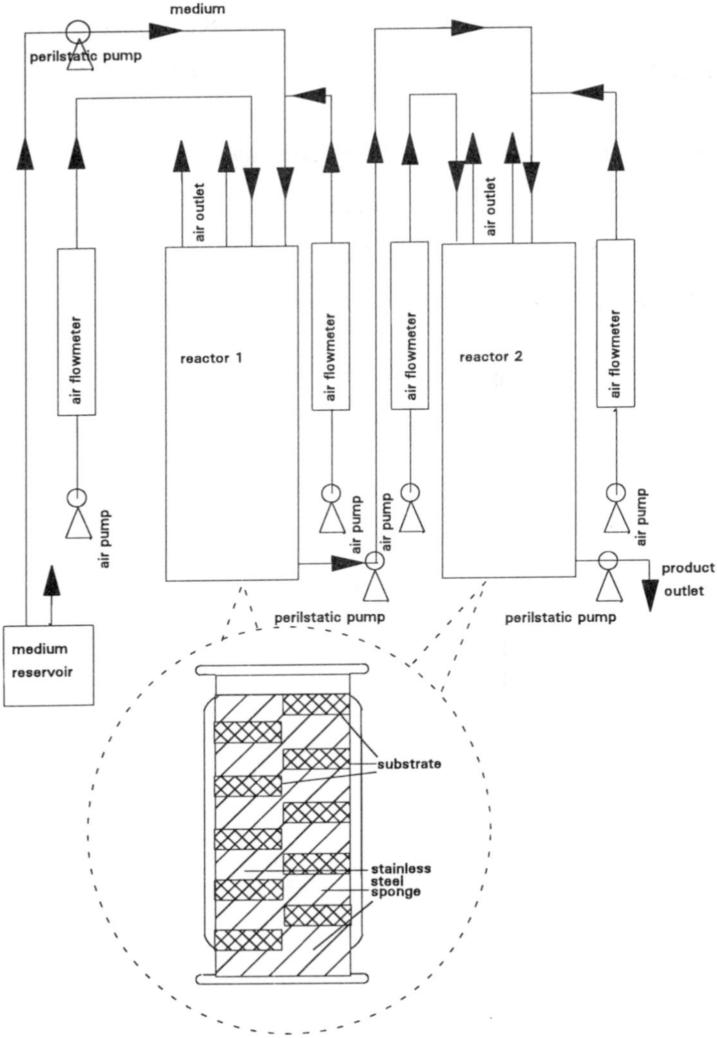

Figure 1: Flow diagram of the reactor set up and the substrate arrangement in the reactor

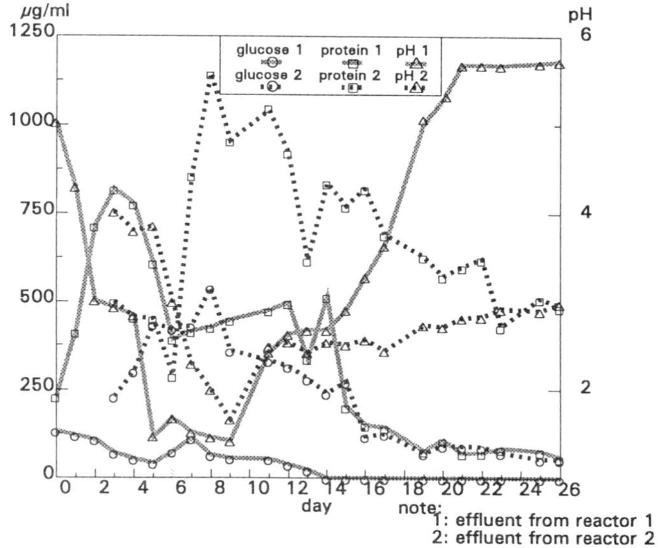

Figure 2: The profile of glucose, protein and pH in the effluent stream of both reactors

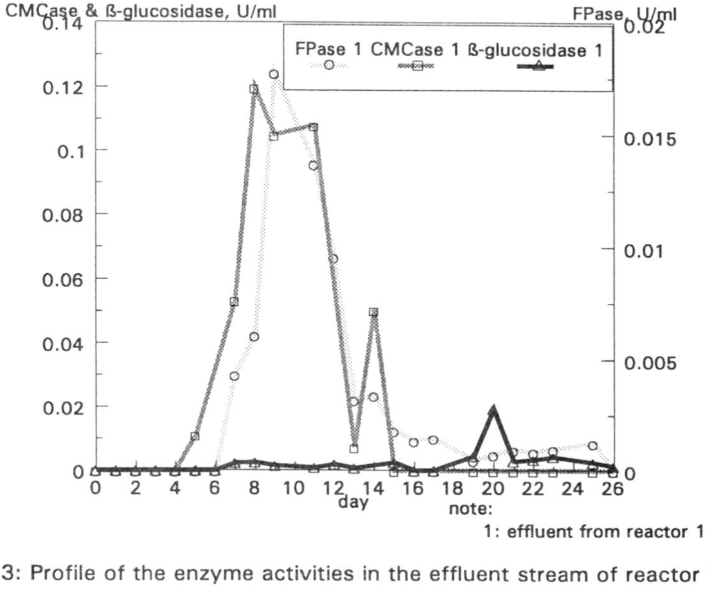

Figure 3: Profile of the enzyme activities in the effluent stream of reactor 1

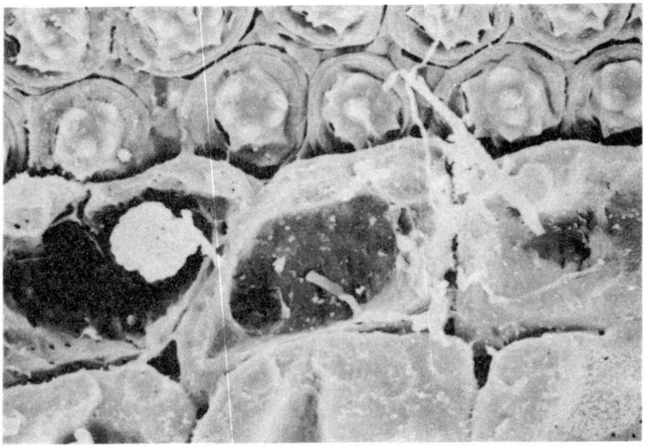

Figure 4: The SEM photomicrograph of substrate from reactor 1 after the enzyme secretion process (x700)

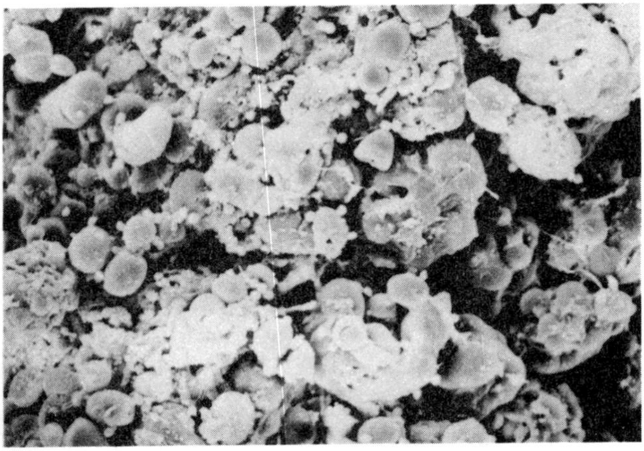

Figure 5: The SEM photomicrograph of substrate from reactor 2 after the saccharification process (x700)

THE PROCESS ENGINEERING OF VEGETABLE SEEDS: PRIMING, DRYING AND COATING

A W Nienow and W Bujalski, BBSRC Centre for Biochemical Engineering
The University of Birmingham, U.K.
R B Maude and D Gray, Horticultural Research International Wellesbourne, U.K.

Work on scaling up the priming, drying and coating of vegetable seeds is discussed. It is shown that the operations can all be conducted successfully in bulk, giving large quantities of seeds with high percentage germination and greatly reduced germination times. The seeds can also be dried rapidly and protected by film coats of either chemical or bacterial biocides. Regardless of whether laboratory or bulk processing methods are used, the useful storage time of primed and dried seeds is not more than about 6 months.

1. INTRODUCTION

This topic has been studied since 1987 in a collaboration between the SERC (now BBSRC) Centre for Biochemical Engineering in the University of Birmingham and HRI, Wellesbourne. The process engineering concept is well-described by the flow sheet of Fig.1. In essence, it suggests a series of operations by which commercial quantities of seed can be treated so that they can be safely stored yet, on sowing, they germinate (and, subsequently, become ready for harvesting) more synchronously and rapidly than untreated seed. In addition, the seeds are protected against a variety of potential pathogens. These are the general agronomic benefits resulting from these bulk treatments and they are of particular relevance to slow germinating species where earliness of production is required. When the work commenced, the priming and drying aspects were established in the laboratory for small quantities of seeds, though drying was very inefficient. Coating was done on a large scale but it too was inefficient. It is the establishment of efficient large scale operation which is discussed here.

2. PRIMING

Priming essentially consists of controlling the water content of seeds so that they imbibe sufficient water to allow the biochemical events preparing the seeds for germination to take place but not actual penetration of the radicle through the seed coat. It is priming which leads, upon sowing, to the advancement of germination.

One of the main methods of priming is to use solutions of controlled osmotic pressure, e.g., polyethylene glycol solutions of molecular weight from 600 to 6000 to give osmotic potentials of the order of -0.5 to -1.5 MPa (Gray et al.,(1)). However, whilst priming is taking place over typically a five to seven day period, the seeds require oxygen to allow the necessary metabolic processes to occur (Nienow and Brocklehurst (2)). On filter paper in Petri dishes, the availability of oxygen depends on the amount of PEG solution used which, if in excess, can cover the surface of the seed, so restricting entry of 0_2. If insufficient solution is available, the concentration of PEG increases simply because seeds imbibe water. Thus, it is difficult to achieve accurate control of osmotic potential which in turn effects seed moisture content and the efficacy of priming(1,Bujalski et al., (3)).

However, the scale-up of seed priming from filter paper to bioreactor technology is very analogous to the scale-up of bioprocesses for other biological species, e.g. fermentation. The supply of oxygen is a particularly critical feature but both it and the osmotic potential can be accurately measured and controlled(3). During priming, pH changes significantly but it does not appear to be necessary to control it in order to achieve satisfactory results(Talavera-Williams et al., (4)). Either bubble column or mechanically agitated bioreactors can be used. They must operate so that all the seeds are fully suspended in the osmotica, thus ensuring that the oxygen and water that they need can be transferred to them. The air must be dispersed into small bubbles for the same reason. Also, like other biological species, there is a level of dissolved oxygen concentrations (dO_2) below which performance deteriorates(Bujalski et al., (5)). Such a level has been shown to be achievable, when using air, for leek(1), carrot(3) and tomato(4). However, for onions, enriched air (air blended with pure oxygen) is required, i.e., higher levels of dO_2 than are achievable with air, to give a measurable reduction in germination times after drying(Bujalski et al., 6)). It is not clear why leek and onion, closely related species biologically, should behave so differently. One possible explanation is that onions have a much thicker seed coat and that this acts as a barrier to oxygen transfer. As a result, the oxygen concentration in the seed is kept below that necessary for priming to be initiated unless the level of oxygen in the liquid is enhanced.

With the other species, e.g.,. leeks, enriched air produces an improved performance in that the treatment time can be reduced by some 30% for the same reduction in germination time. For example, when sparging with nitrogen to give zero $d0_2$, a reduction in germination time is not achieved. With increasing levels of $d0_2$ as a result of raising the concentration of 0_2 in the nitrogen, the germination time is progressively reduced, roughly in proportion to the increase in $d0_2$. This reduction continues up to oxygen concentrations equivalent to air. After that the effect progressively lessens until at about 50:50 oxygen/nitrogen mixture, no further reduction in germination time is achieved. Thus, the priming process fits a Michaelis-Menton model in accordance with the enzymatic processes which take place during priming(5). This model would also support the explanation given for the difference in behaviour between onions and leeks.

In bubble columns, the air flow acts as both the means of seed suspension as well as the provider of oxygen and this configuration is often the cheapest option. However, though stirred bioreactors are more complex and expensive, there is increased flexibility. Thus, they are preferable for commercial scale priming if the use of expensive enriched air is contemplated since agitation achieves seed suspension and only low gas flow rates are required to satisfy the oxygen demand and $d0_2$ level. However, the choice of agitation conditions has to be made very carefully, otherwise the seeds are damaged during priming(2). Choosing conditions which achieve seed suspension under aerated conditions with minimum mixing intensity can eliminate this problem(2).

Other important considerations for successful commercialisation are seed concentration and re-use of PEG. Seeds have been successfully primed in stirred bioreactors up to 50 litres at concentrations up to 100g seed per litre of PEG, though, at such concentrations, vortex reactors (Fig.2) are preferable to bubble columns if agitators are not used. It has also been shown that

PEG can be re-used at least twice without a damaging build-up in bacterial and fungal microflora or any deterioration in priming performance(Petch et al., (7)).

Overall, typical reductions in germination time for freshly primed seeds are from 3.5 days to as little as 0.5 days for leeks(1); from 3.6 to 1.6 for tomatoes(4); from 6.0 to 1.6 for onions(6); and from 5.6 to 1.5 for carrots(3). In all cases, results as good as or better than priming on filter paper were achieved.

3. DRYING

Drying of seeds after priming increases their mean germination time compared to undried seeds by about 1 day because of the need to re-imbibe water(3). Large quantities of seeds have been dried in air-fluidised beds operated using ambient air or air heated up to 30 to 40°C (though the bed is kept much cooler). Mass and heat transfer considerations predict a drying time of about 30 min to an hour and such times have been achieved (4, Bujalski et al., (8)). This time should be compared to the 16h or more commonly used in conventional laboratory thin layer tray dryers. These fluidised bed-dried seeds give germination and seedling emergence performance at least as good as controls based on filter paper priming and thin-layer drying. Thus, much more rapid moisture removal can be used without loss of seed viability than had been believed to be the case. However, seed from either slow laboratory treatment drying or the faster fluidised bed route (whether primed osmotically or by other possible methods such as drum priming(Gray et al., (9)) start to show an increase in abnormal seedlings after some six months, even though germination remains the same as in controls; and this deterioration is progressive(Maude et al., (10)). In order to show conclusively this deterioration, independent of scale of operation, a new seedling assessment technique has had to be developed (10).

4. FLUIDISED BED COATING

This technique (called in the seed trade, film coating) was carried out in the same bed as that for drying, using a modified two-fluid, atomising nozzle(Maroglou and Nienow (11)). Film coating provides a very efficient means of applying chemical biocides (Nienow (12)). Compared to other techniques, the dosage on each seed is much more uniform and is stable during subsequent handling. Thus, a much lower average dosage can be used which is highly desirable environmentally. Interestingly, in spite of the intense shear stress, shear rate and two phase atomised flow produced in such nozzles, bacterial biocides have been shown to remain viable in seed coats; and this viability has been maintained for at least 6 months (12).

5. THE OVERALL PROCESS

Leek(8) and naturally infected carrots(Maude et al., (13)) have been given the complete treatment indicated by Fig.1. In both cases, laboratory tests and glasshouse and field trials have given excellent germination advance and seedling emergence characteristics; and in the case of carrots, disease control(13). Again, abnormal seedlings increase from 6 months storage onwards(10).

6. CONCLUSION

This work has clearly shown the viability of the large scale process engineering of seeds. However, it has also shown that if the seeds are well primed, the useful storage time of dried seeds is limited to about 6 months, regardless of the scale at which the process is conducted.

7. ACKNOWLEDGEMENT

The authors thank the AFRC for financial support for this work.

8. REFERENCES

1. Gray, D., Drew, R.L.K., Bujalski, W. and Nienow, A.W. (1991), "Comparison of Polyethylene Glycol Polymers, Betaine and L-proline for Seed Priming", Seed Sci. and Technol., 19, 581.
2. Nienow, A.W. and Brocklehurst, P.A. (1987), "Seed Preparation for Rapid Germination - Engineering Studies", in "Bioreactors and Biotransformations". (Ed. G.W. Moody and P.B. Baker), Elsevier, 1987, pp 52-63.
3. Bujalski, W., Nienow, A.W., Petch, G.M. and Gray, D. (1991), "Scale-up Studies for Osmotic Priming and Drying of Carrot Seeds", J. Agric. Engng.Res., 48, 287.
4. Talavera-Williams, C.G., Pacek, A.W., Bujalski, W. and Nienow, A.W. (1991), "A Feasibility Study of the Bulk Priming and Drying of Tomato Seeds", Trans I Chem E, part C, 69, 134.
5. Bujalski, W., Nienow, A.W., Maude, R.B. and Gray D. (1993), "Priming Responses of Leek (*Allium porrum* L.) Seeds to Different Dissolved Oxygen Levels in the Osmoticum" Ann. Appl. Biol., 122, 569-577.
6. Bujalski, W., Nienow, A.W. and Gray, D. (1989), "Establishing the Large Scale Osmotic Priming of Onion Seeds by Using Enriched Air", Ann. Appl. Biol., 115, 171.
7. Petch, G.M., Maude, R.B., Bujalski, W. and Nienow, A.W. (1991), "The Effects of Re-use of Polyethylene Glycol Priming Osmotica Upon the Development of Microbial Population and Germination of Leeks and Carrots", Ann. Appl. Biol., 119, 365.
8. Bujalski, W., Nienow, A.W., Petch, G.M., Drew, R.L.K. and Maude, R.B. (1992), "The Process Engineering of Leek Seeds: a Feasibility Study", Seed Sci. and Technol., 20, 129.
9. Gray, D., Rowse H.R., Finch-Savage W.E., Bujalski W. and Nienow A.W. (1993), "Priming of Seeds - Scaling up for Commercial Use", Proc. 4th Int. Workshop on Seeds; Basic and Applied Aspects of Seed Biology; Vol.3, (D. Côme and F. Contineau, Eds.) Université Pierre and Marie Curie, Paris, France, pp 927-934.
10. Maude, R.B., Drew, R.L.K., Gray D., Bujalski, W. and Nienow, A.W. (1993), "The Effect of Priming on the Storage of Leek Seeds", Seed Sci. and Technol. (in press).
11. Maroglou, A. and Nienow, A.W. (1985), "Mechanisms of Particle Growth in Fluidised Bed Granulation", Proc. 4th Int. Symp. on Agglomeration; Iron and Steel Society Inc., Toronto. pp 465-470.
12. Nienow, A.W. (1993), "Fluidised Bed Granulation and Coating: Applications to Materials, Agriculture and Biotechnology", Proc. 6th Int. Symp. on Agglomeration, Nagoya, Japan, pp 1-10.

13. Maude, R.B., Drew, R.L.K., Gray, D., Petch, G.M., Bujalski, W. and Nienow, A.W. (1992), "Strategies for Control of Seed Born *Alternaria dauci* (Leaf Blight) of Carrots in Priming and Process Engineering Systems", Plant Pathology, 41, 204.

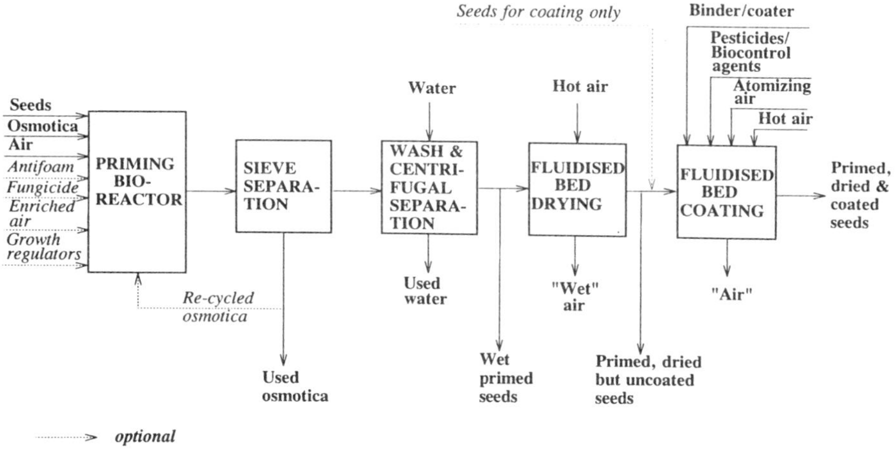

Figure 1. A flow sheet for the process engineering of vegetable seeds.

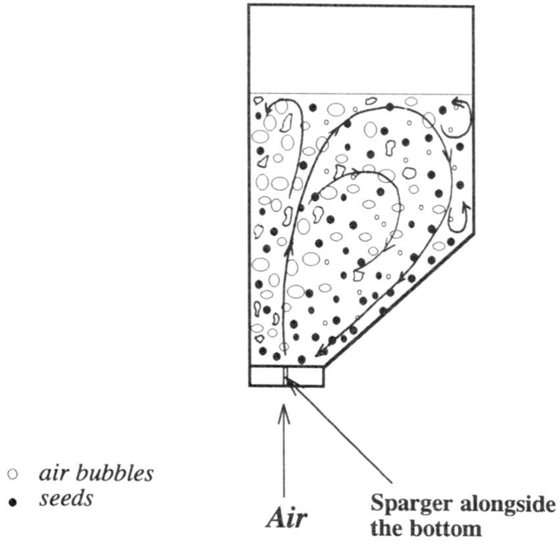

Figure 2. A"Vortex" Bulk Priming Bioreactor.

BIOREMEDIATION OF PROCESS WATERS CONTAMINATED WITH SELENIUM

L. Riadi and J.P. Barford
Department of Chemical Engineering, University of Sydney, NSW 2006, Australia.

Samples were collected from a coal combustion disposal site. This ash dam was a known source of selenium pollution. The samples were screened for naturally bacteria present. Some organisms (SP1, *Pseudomonas pickettii, Arthrobacter protophormeae, Bacillus pasteurii, Staphylococcus haemolyticus*, strain AD-1) isolated from the site were capable of reducing selenite to elemental selenium. The results indicated that one of the organisms (SP1) has potential to remove selenite rapidly. We have used this organism to develop and demonstrate a novel bioremediation process for selenium.

1. INTRODUCTION

Selenium, once of major concern as a toxic element has four possible oxidation states (Se^{6+}, Se^{4+}, Se^{0}, Se^{2-}). Selenate (SeO_4^{2-}) and selenite (SeO_3^{2-}), the dominant forms of selenium in water are toxic and soluble whereas elemental selenium is non-toxic, highly insoluble in water and very stable. The most reduced form of selenium is selenide (Se^{2-}) which can be present as gas and is highly toxic. However, it is seldom a threat since it is rapidly oxidised to insoluble elemental selenium in the presence of air. The recommended standards for selenium in drinking water and for irrigation are 0.01 mg and 0.05 mg respectively (11). Many organisms have been reported to reduce selenium oxides (2, 3, 6, 7, 8, 9, 10, 11), and some of them (8, 11) have shown that selenate or selenite was reduced anaerobically.

The identification of high-level resistance (HLR) to selenite and selenate in ten isolates from the site has indicated the possibility that these organisms could be used to reduced these compounds when grown under oxygen-limited condition. The predominant species of selenium in the contaminated site is selenite and the level of selenate is less than the allowable discharge concentration. Therefore, it is obvious that an effort is needed to overcome the unacceptable level of selenite on site. The use of indigenous microorganisms to solve this environmental problem has been investigated. In this paper, we present Minimum Inhibition Concentrations (MIC), possible organic carbon sources that exist on site, and fundamental kinetic data of these organisms.

2. MATERIALS AND METHODS

2.1 Sampling and Isolation of Organisms

Samples were obtained from process waters (overflow stream, recycle stream and underflow stream) and the sediment cores (50 cm x 6 cm) from an ash dam on the contaminated site. Grab samples were placed in sterile containers and stored at 4°C and - 20°C and processed within 24 hours of collection. The pH values for overflow stream, recycle stream and underflow stream were 8.0, 7.5 and 7.0, respectively. Upon collection, serial tenfold dilutions of homogenised samples were made in 0.1% sterile peptone diluent. Samples (100 μL) of each dilution were spread on nutrient agar media (Oxoid). The enrichment study was performed on nutrient agar media supplemented with 200 μg/L selenium as sodium selenate or sodium selenite. Selenium salts added to media were from fresh, filter-sterilised solutions. Plates were incubated at room temperature for 4-5 days.

2.2 Identification of Organisms, Growth Conditions and Media

Individual colony taken from these plates and identified by standard microbiological techniques (5) to genus level, and using gas liquid chromatography to identify their component fatty acids and comparing these against the anaerobe library database for genus and species level. The fatty acids identification was performed by biological and chemical Research Institute of NSW Agriculture Department.

Two repetitions of each experiment were conducted with cultures which exhibited the ability to reduce selenite. A 10 % log phase cultures growing in nutrient broth media (yeast extract 2 g/L, peptone 5 g/L, Lab Lemco 1 g/L, NaCl 5 g/L) was transferred to media containing 200 μg/L of selenite. Inoculated media were incubated statically at 27°C, pH 7.2. During culture growth, samples were withdrawn periodically and analysed for residual selenium concentration and microbial growth (estimated by dry weight (5)). An additional control experiment consisting of heat-killed cells (heat at 100°C for 15 minutes) was also performed.

2.3 Minimum Inhibition Concentration (MIC)

MICs of selenite and selenate were performed by tube dilution method to determine the relative toxicity of selenium salts toward these organisms. Cultures to be tested were grown to late log phase in nutrient broth media. Serial twofold dilutions of the metals were made in nutrient broth in test tubes. A 10 % inoculum of the bacteria was added and the tubes were incubated. The MIC was defined as the lowest concentration of inhibitor preventing growth at 27°C after 48 h of incubation for SP1 and *Pseudomonas pickettii* and 72 h for other species.

2.4 Analysis

Selenium. Selenium oxyanion concentrations were determined by flow-through hydride generation atomic absorption spectrometry (196 nm). A closed-system HCl reduction (10 M, 100 °C, 15 minutes) of selenate and selenite was employed for determinations of Se(IV) plus Se(VI) (4). Selenate concentrations were derived by the difference in selenite concentrations of the HCl-treated and untreated samples. Detection limits for selenium was 0.3 μg/L. Dried red precipitate from cultures was analysed using a Philips X-ray diffraction unit model 1130.

Acetate, VFA, COD. Acetate was determined using the acetic acid UV test kit of Boehringer Manheim.

Volatile fatty acids concentrations were determined by absorption of water sample on a silicic acid column using chloroform-butanol reagent and titration of the aliquot using 0.02 M NaOH (1).

Chemical Oxygen Demand was determined by cell test photometer, 446 nm wavelength. The samples were added to a reaction cell containing reagent mixture from Merck and heated in a thermo-reactor (TR 205, Merck) for 120 minutes at 148°C prior to determination.

3. RESULTS AND DISCUSSIONS

3.1 Identification of Organism

The isolates were a diverse group, including five gram positive isolates and fifteen gram negative isolates. After an enrichment study, ten out of twenty isolates were found either resistant to selenite or selenate. These isolates were SP1, *Pseudomonas pickettii, Staphylococcus haemolyticus, Bacillus pasteurii, Methylobacterium radiotolerens, Staphylococcus epidermidis, Pseudomonas saccarophila,* strain AD-1 (gram positive, catalase positive, oxidase positive with budding shape), Strain AD-2 (gram negative rod, catalase positive, oxidase positive).

3.2 MIC

The MICs of sodium selenite (Table 1) varied from 6.25 mM to over 200 mM whereas of sodium selenate ranged from 50 mM to over 400 mM. From the data obtained, it can be seen that the relative toxicity of selenite and selenate varied for each strain. Six isolates (SP1, *Arthrobacter protophormeae, Pseudomonas pickettii, Strain AD-1, Staphylococcus haemolyticus* and *Bacillus pasteurii*) produced red color in the presence of selenite. After detection using X-ray diffraction, the red color was shown the amorphous red Se^0 (s). None of the isolates produced red color when exposed to selenate.

As illustrated from Table 1, there is significant bacterial resistance to selenite and selenate in a selenium-contaminated aquatic system.

3.3 Selenium Level and Organic Carbon on Site

The selenium level measured on site is presented in Table 2. It can be seen that selenite is the predominant species compared to selenate.

The determination of total and/or individual organic carbon species (Table 3) will give an insight into what is the likely carbon source on site as well as assisting in postulating the

Table 1. MICs of selenite and selenate for bacterial isolates from an ash dam site.

ISOLATE	MIC (mM) for :	
	Se(IV)	Se(VI)
SP1	over 200	over 400
Pseudomonas pickettii.	200	400
Arthrobacter protophormeae	50	100
Staphylococcus haemelyticus	50	400
Bacillus pasteurii	100	400
Staphylococcus epidermidis	12.5	200
Pseudomonas saccarophila	6.25	100
Methylobacterium radiotolerans	6.25	200
Strain AD-1	10	200
Strain AD-2	10	50

likely biochemical pathways used for their metabolism. The COD level represented the amount of organic carbon on site. Lactate was not found in the water and sediment samples. Acetate was present and presumably arose from urban runoff. Theoretically, 1 g/L acetate is equivalent to 533 mg/L COD, hence, the amount of acetate in water samples is not sufficient enough to represent the organic carbon on site. However, the amount of acetate in the sediment is approximates the COD measured.

Table 2. Selenium levels on site.

Sample	μg/L		
	Total Se	Se(VI)	Se(IV)
Overflow stream	89	13	76
Recycle stream	121	46	75
Underflow stream	22	-	-

Table 3. COD, VFA and Acetate on site.

Sample	mg/L		
	COD	Acetate	VFA
Overflow stream	300	3.8	36
Recycle stream	350	2.27	20
Underflow stream	650	2	110
Sediment	250	393	

3.4 Growth of Isolates in the Presence of Selenite

There was rapid and complete removal of selenite by SP1 and *Pseudomonas pickettii* (Fig 1 and 2). Analysis revealed that all selenite had been taken up by SP1 and *Pseudomonas pickettii* after 3 and 8 hours, respectively. However, other organisms (*Staphylococcus haemolyticus, Bacillus pasteurii, Arthrobacter prothophormeae*) did not remove selenite completely. The significant difference of uptake of selenite by live cells and heat-killed cells demonstrated an active biological transformation (Fig. 1, 2, 3, 4 and 5). In strain AD-1, the amount of selenite taken up by live cells and heat-killed cells was almost the same (see Fig 6). This uptake was predominantly due to adsorption on the cell wall, but active biological transformation also occurred. Oxygen-limited conditions were used in the experiment to avoid the inhibition of selenite reduction by oxygen as a competitive electron acceptor. The fundamental kinetic data such as specific growth rate, specific uptake rate and yield of these organisms (see Table 4) are calculated according to these equations (X = biomass):

$$\mu = \frac{1}{X}\frac{dX}{dt} \tag{1}$$

$$Y_{xs} = \frac{dX}{dSe} \tag{2}$$

$$Q_s = \frac{\mu}{Y_{xs}} \tag{3}$$

From the fundamental kinetic data, it can be seen that among the organisms SP1 has the highest specific growth rate and selenium uptake rate.

Table 4. Specific growth rate (μ), Specific uptake rate (Qs), efficiency of selenium reduction and Yield (Yxs) of the organisms isolated from site.

Isolates	μ (h^{-1})	Qs (μg Se/g cell)	% Se reduction	Yxs g cell/mg Se
SP1	0.24	2474	100	0.1
Pseudomonas pickettii	0.23	940	98.3	0.25
Staphylococcus haemolyticus	0.021	26	90.1	0.81
Arthrobacter protophormeae	0.025	8.1	10.53	3.15
Bacillus pasteurii	0.057	28.2	29	2
Strain AD-1.	0.027	2.02	7.85	13.4

4.0 CONCLUSIONS

These results indicate that selenite respiration by SP1 may allow the development of an biological selenium removal process for selenium-contaminated aquatic systems based on forming elemental selenium by bacterial reduction. Acetate can be considered as a posiible carbon sources for this process which involves biological degradation by the TCA cycle. However, other carbon sources need to be assesed as possible practical alternatives.

5.0 ACKNOWLEDGEMENTS

We thank to Dr. Gaby Deleon for his help in X-ray diffraction.

6.0 REFERENCES

1. American Public Health Association, 1976, Standard methods for the examination of water and wastewater, 14th ed., p 527.
2. Baldwin, R.A., Stauter, J.C., Kauffman, E. and Laughlin, W.C., 1985, US Patent No. 4519913.
3. Burton, G.A. Jr., Giddings, T.H., DeBrine, P. and Fall, R., 1987, App. Environ. Microbiol., 53, 185-188.
4. Bye, R. and Lund, W., 1988, Fresenius Z. Anal. Chem., 332, 242-244.
5. Gerhardt, P., Murray, R.G.E., Costilow, R.N., Nester, E.W., Wood, W.A., Krieg, N.R. and Phillips, G.B., 1981, Manual of methods for general bacteriology, American Society for Microbiology, Washington D.C., 411-443.
6. Gersberg, R.M., 1986, Report CATI/80621, California State University.
7. Larsen, D.M., Gardner, K.R., Altringer, P.B., 1989, SME & AIME, 177-185.
8. Macy, J.M., Michel, T.A. and Kirsch, D.G., 1989, FEMS Microbiol. Letts., 61, 195-198.
9. Maiers, D.T., Wilchlacz, P.L., Thompson, D.L. and Bruhn, D.F., 1988, App. Environ Microbiol., 54, 2591-2593.
10. McCready, R.G.L., Campbell, J.N. and Payne, J.I., 1966, Can. J. Microbiol., 12, 703-713.
11. Oremland R.S., Hollibaugh, J.T., Maest, A.S., Presser, T.S., Miller, L.G. and Culberston, C.W., 1989, App.Environ.Microbiol., 55, 9, 2333-2343.

SYMBOLS

HLR	high-level resistance
MIC	Minimum inhibition concentration
TSBA	Trypticase soy broth agar
VFA	Volatile fatty acids
COD	Chemical oxygen demand
μ	specific growth rate
Qs	specific uptake rate
Yxs	yield coefficient

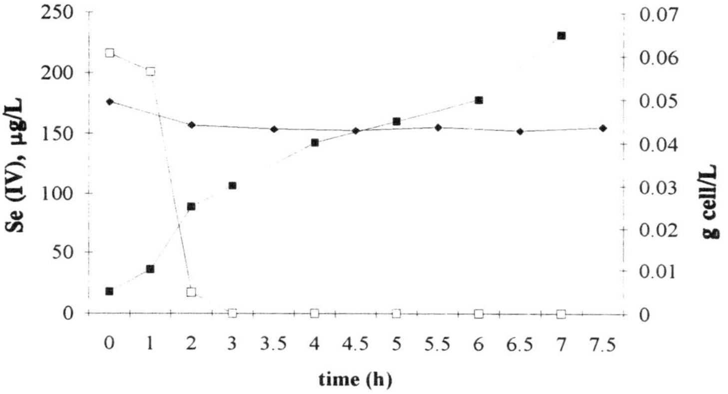

Fig.1. Effect of time on growth of SP1 (■) and removal of selenite in live cells (□) and in heat-killed control (◆).

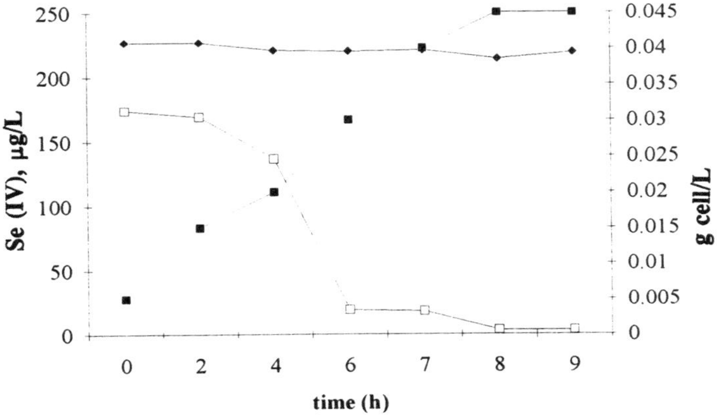

Fig.2. Effect of time on growth of *Pseudomonas pickettii* (■) and removal of selenite in live cells (□) and in heat killed control (◆) .

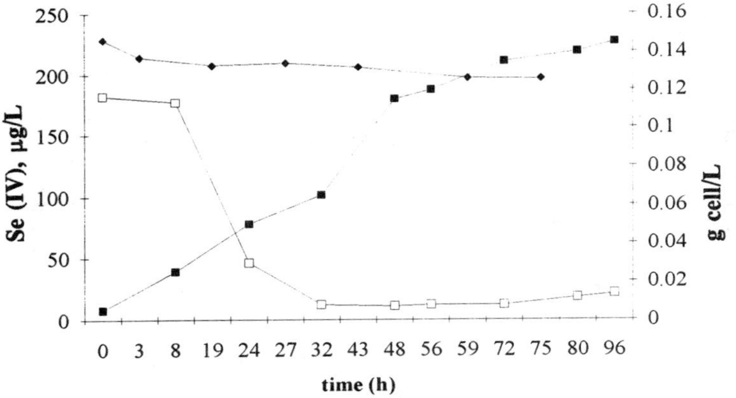

Fig.3. Effect of time on growth of *Staphylococcus haemolyticus*(■) and removal of selenite in live cells (□) and in heat killed control (◆).

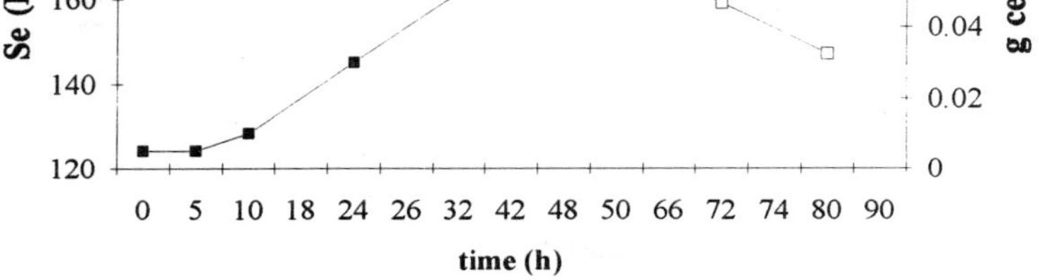

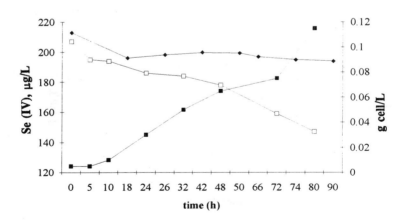

Fig.4. Effect of time on growth of *Bacillus pasteurii* (■) and removal of selenite in live cells (□) and in heat killed control (◆) .

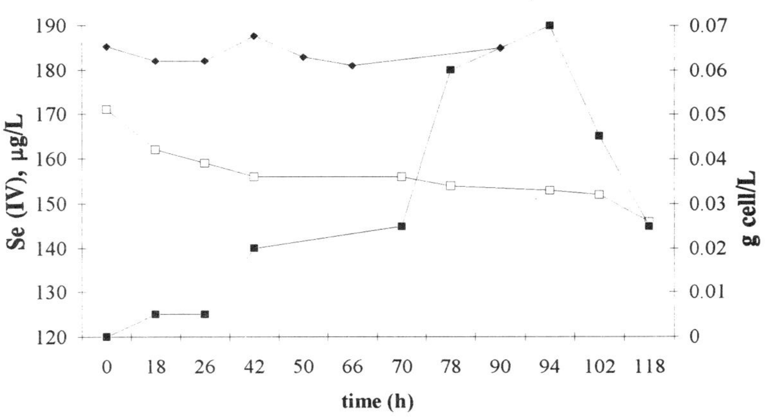

Fig.5. Effect of time on growth of *Arthrobacter protophormeae*(■) and removal of selenite in live cells (□) and in heat killed control (◆).

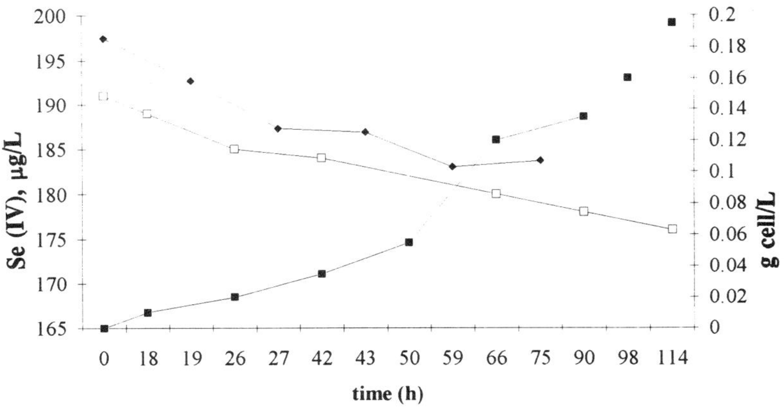

Fig.6. Effect of time on growth of Strain AD-1(■) and removal of selenite in live cells (□) and in heat killed control (◆).

AN AQUEOUS TWO-PHASE PARTITIONING SYSTEM OF PEG-ENZYME AND TANNIC ACID FOR RECOVERY AND REUSE OF CELLULASE IN PALM WASTE SACCHARIFICATION

K. Das* , G. Dutta and M. A. Amiza, Division of Food Technology, School of Industrial Technology, Universiti Sains Malaysia, 11800 Penang, MALAYSIA
* Corresponding Author

The application of an aqueous two-phase partitioning system for the recovery and reuse of enzyme was studied. By addition of polyethylene glycol (PEG) to a complex of tannic acid (TA) and cellulase, a two-phase system was developed with liberation of the enzyme in solution for reuse in the saccharification of palm press fibres (PPF). Investigation of the phase separation and enzyme recovery indicated that the enzyme concentration required was 3% (v/v) which means that its concentration should be around 8% (w/w) during hydrolysis before dilution with TA and PEG. The conditions required for the formation of two-phase systems include a pH range of 4.0-5.5, temperature of 30-45°C and addition of NaCl (0.5M). The conditions conducive for maximum recovery were pH 5.5 and a temperature of 45°C. It was also observed from the results that TA may partially inhibit enzyme activity.

1. INTRODUCTION

Recovery and reuse of enzymes after hydrolysis may be achieved by the use of membranes as shown by Anis *et al* (1) or by immobilisation of soluble dextran done by Tjerneld *et al* (2). A newer method for extractive bioconversion and recovery of enzyme based on a two-phase aqueous system formed by mixing aqueous solutions of two water-soluble polymers may be promising.

Two-phase phenol-water system has limited use since phenol may denature protein. The possibilities for the use of different aqueous two phase systems and their classification have been elaborated by Albertsson (3). The present study was designed to develop a possible process for the formation of an aqueous two-phase system by addition of PEG to TA-cellulase complex to release the enzyme for continuous recycle and reuse in saccharification.

2. MATERIALS AND METHODS

Palm press fibres obtained from Malpom Industries, Penang, Novo cellulase with an activity of 1500 IU/ml and PEG and TA from BDH were used in this study. Assays include the DNS method after Miller (4) for reducing sugars, Folin-Lowry method of Peterson (5) for protein, the CMC-method for cellulase activity and spectrophotometer (295 nm) for TA.

Binodial curves were drawn following studies of a series of two-phase compositions of PEG and cellulase. The experimental set-up for phase separation and enzyme recovery is shown in Figure 1.

3. RESULTS AND DISCUSSION

3.1 Effects of pH, Temperature and Salt on the PEG-enzyme Two-phase Formation

As seen in Figure 2, with increase in pH, the binodial curves move away from both the X- and Y- axes indicating increases in both cellulase and PEG concentrations needed for phase separation. At pH 4 and 5, the respective enzyme concentrations were 1 and 2.5% (v/v), i.e. 15 and 37.5 U/ml, when the PEG concentration was 14% (w/v). This means that at lower pH, phase separation occurs at a lower concentration of enzyme, which is advantageous since a low enzyme concentration is generally used during saccharification. But with increase in temperature (Figure 3), it was found that higher enzyme concentrations were required for the two-phase formation, though differences between the binodials at different temperatures were not large. In fact, pH played a more important role than temperature in phase separation. The experimental results (Figures 2 and 4) also show that with addition of NaCl, the binodial curves moved closer to both axes resulting in lower concentrations of enzyme for phase separation, as also pointed out by Albertsson (3).

3.2 Effect of pH on Partitioning Process for Enzyme Recovery

The results shown in Figure 5 indicate that an increase in pH increased the TA concentration within the PEG-phase, i.e. more TA was liberated from the enzyme-TA complex as the pH was increased.

3.3 Effect of Temperature on Partitioning Process for Enzyme Recovery

Figure 5 shows that the concentration of TA in the PEG phase increased as the temperature was increased from 30 to 45°C, with the partitioning of PEG reaching a maximum at 45°C and pH 5.5. At temperatures above 45°C, the concentration of TA in the PEG phase declined at all pH values.

4. Effect of Tannic Acid on Enzyme Recovery and Activity

4.1 Without Partitioning (when TA Content within the Enzyme Solution was Relatively High)

The relative specific activity of the enzyme and the extent of its recovery at various concentrations of TA are shown in Figure 6. The results indicate that without

TA , the amount of enzyme recovered was 64% and the relative specific activity was 96%, which means that the amount of active enzyme recovered was actually 61%. With additions of 0.5, 1, 2, 3 and 4% TA, the recoveries were 67, 69, 73, 76 and 78% respectively, and the relative specific activities are 88, 85, 78, 73 and 68 %. This shows that TA increased enzyme recovery, but reduced its activity. Ohba *et al* (6) observed a similar pattern for pullulanase. This may be due to steric effects occurring when TA forms hydrogen bonds with the enzyme, thus blocking its active sites for catalytic action.

4.2 With Partioning

Figure 7 shows the amounts of enzyme recovered and the relative specific activity of the enzyme after partitioning at various concentrations of TA. From the results, it is seen that the relative specific activity is lower than that before partitioning (Figure 6), but there is an increase in enzyme recovery from 64 to 69%. With increases in the concentration of TA from 0 to 30%, the amount of enzyme activity recovered increased from 61% to 65%. But at a concentration of 4% TA, the amount of enzyme activity recovered decreased to 60% . This might be due to high concentrations of TA disturbing the equilibrium of the system and/or to the high concentration in the enzyme phase reducing the specific activity of the enzyme. However, the amount of active enzyme recovered was higher than that without partitioning, even in this case.

ACKNOWLEDGEMENTS

The authors are grateful to 'Malayan Sugar Manufacturing (MSM) Co. Bhd.' for their financial support in this work, and to Cik Anis M and Cik Astimar A.A for their typographical assistance.

REFERENCES

1. Anis, M, Das, K and Ismail, N, 1990. J. Biosci., 1(2), 151
2. Tjerneld, F, Persson, I, Albertsson, P and Mahn-Magerdal, B,1985. Biotech. Bioengg., XXVII, 1036
3. Albertsson, P. A , 1971. Partition of Cell Particles and Macro- molecules, John Wiley and Sons, New York
4. Miller, G. L , 1959. Anal. Chem., 31, 426
5. Peterson, G. L , 1979 . Anal. Biochem., 100, 201
6. Ohba, R, Chean, H, Hayashi, S, and Uedas ,1987. Biotechnol. Bioeng., 20, 665

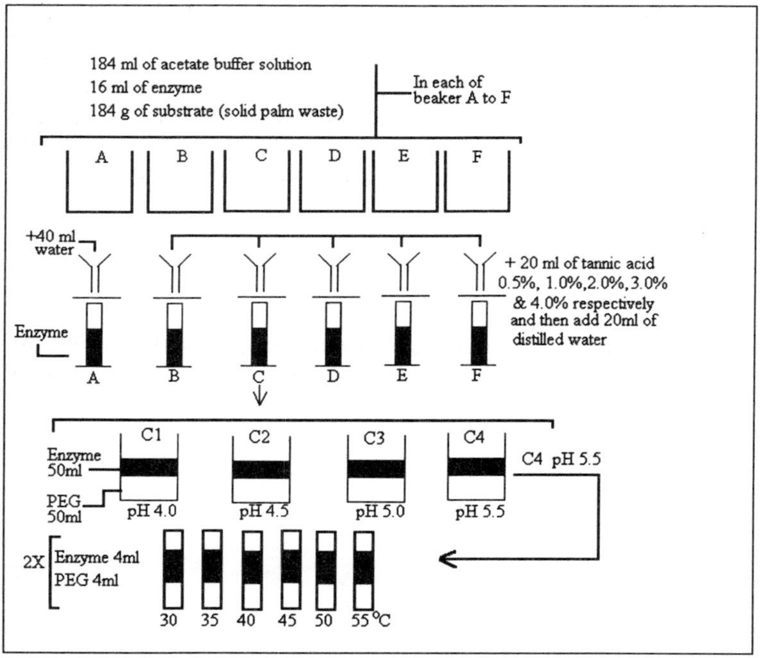

Fig. 1. Phase separation and the recovery process of the enzyme

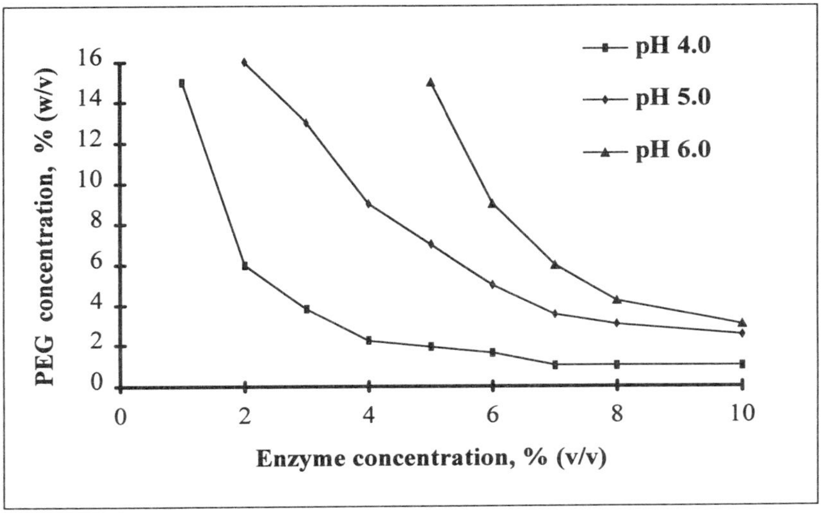

Fig. 2. Effect of pH on phase formation of PEG-cellulase system (concentration of NaCl 0.5M)

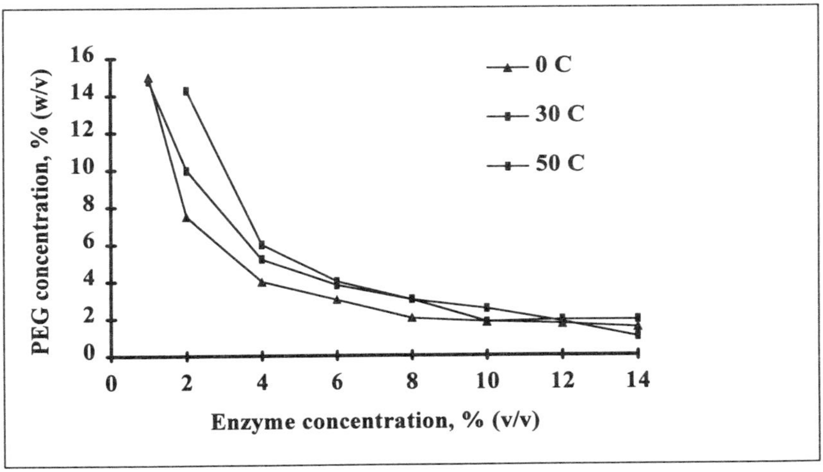

Fig. 3. Binodial curves for PEG-cellulase system at different temperatures (pH 4.0, concentration of NaCl 0.5M)

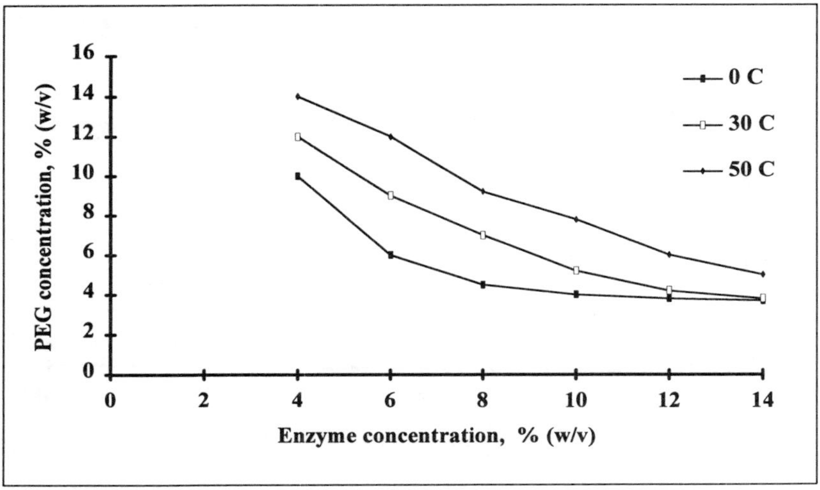

Fig. 4. Binodial curves for PEG-cellulase system at different temperatures (pH 4.0, no NaCl)

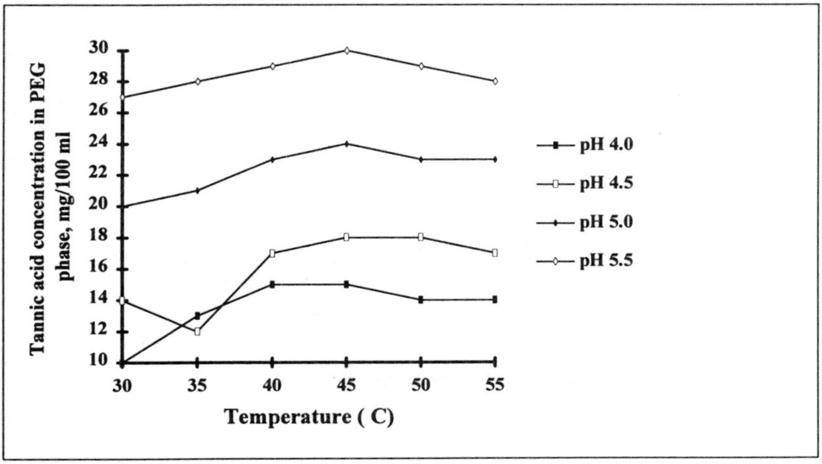

Fig. 5. The correlation between TA concentration in the PEG-phase and pH at different temperatures. (TA used, 0.5%; NaCl concentration 0.5 M)

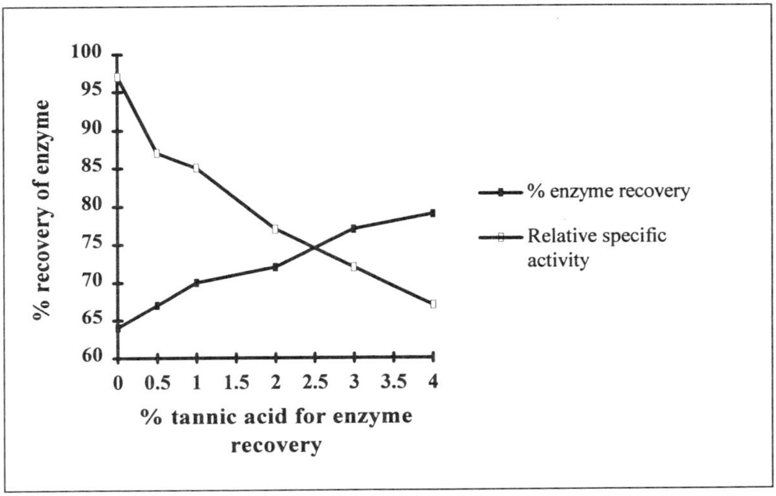

Fig. 6. Relationship between % recovery of enzyme and % TA used for its recovery with relative specific activity before adding PEG (temperature of 45°C, pH 5.5)

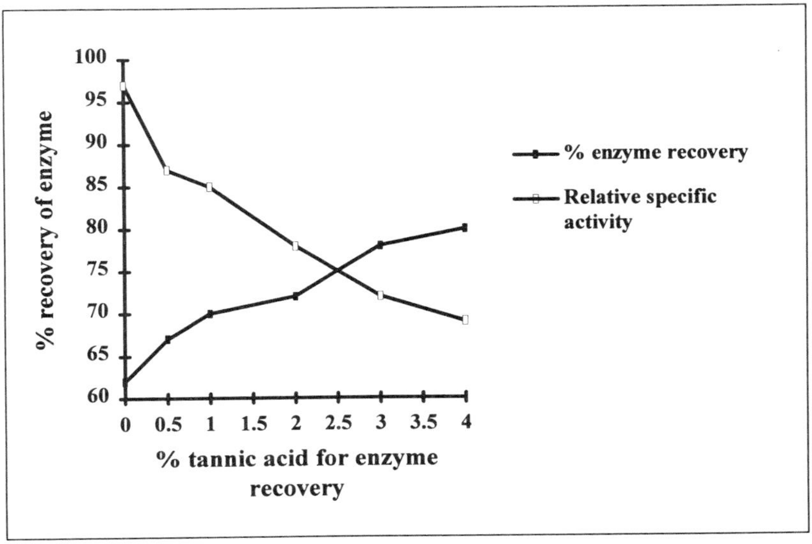

Fig. 7. Relationship between % recovery of enzyme and TA used for its recovery with relative specific activity after adding PEG. (temperature of 45°C; pH 5.5)

COMPUTER CONTROL OF FERMENTATION PROCESSES BY AI TECHNOLOGIES

Toshiomi Yoshida
International Center of Cooperative Research in Biotechnology, Japan, Faculty of Engineering, Osaka University, 2–1, Yamada–oka, Suita, Osaka 565 Japan

The concept of physiological state control of bioreactor systems is introduced along with extensive discussion on the characteristics of bioreactor control systems. A software system for the diagnosis of the physiological situation through the inference procedure utilizing a fuzzy set approach and a physiological state control system in a hierarchy structure are demonstrated. The control system was applied on the fed–batch cuolture for phenylalanine production by a recombinant *Escherichia coli*. The investigation on the detection and control of physiological state variability in the continuous culture of the organisms is also demonstrated.

1. INTRODUCTION

During the last two decades the intensive research has been made on the automatic control of bioreactor systems. Although the conventional control methods are still deeply entrenched in both the theory and practice, it is becoming clear that alternative approaches, better suited to the nature of living systems, have to be developed(1–3). Recently, there is a tendency towards interpretation of the problem for control of bioreactor systems from informal viewpoints, using nontraditional, artificial intelligence (AI) techniques or knowledge engineering (KE) approaches(4–11). This paper introduces the concept of the physiological state control, explains the classification of various kinds of biologically representative variables, and discusses our recent works in developing new approaches for identification of the state of biological processes, especially the physiological state of cells, and also the detection and control of physiological state transition in fed–batch and continuous culture for production of phenylalanine by a recombinant *Escherichia coli*.

2. DEVELOPMENT OF ARTIFICIAL INTELLIGENCE

The main purpose of the study on artificial intelligence (AI) has been the building of a more intelligent machine to imitate the human thinking. In this context, the machine is a computer, and the purpose will be fulfilled by development of programming; the true nature of AI would be the investigation on development of an intelligent computer or an intelligent program. AI is a sub–field of computer science, and is concerned primarily with symbolic reasoning and problem solving. Today the science of AI spans a growing list of emerging disciplines: knowledge representation, problem solving, expert and knowledge systems, learning, natural language interfacing, cognitive modeling, and robotics (Figure 1).

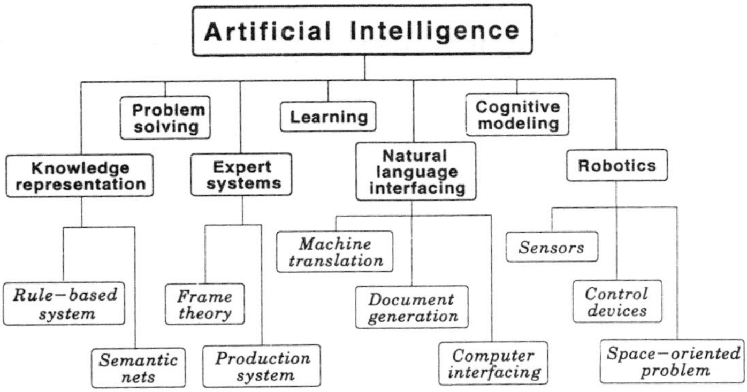

Figure 1. Emerging disciplines in Artificial Intelligence science.

2.1. Knowledge Engineering

A problem-solving intelligent machine has not been realized by seeking "a simple and general purpose inference mechanism" which early AI researchers dreamed of, while the purpose seems to be fulfilled by "utilization of a large quantity of knowledge in a particular domain." The development of this new stream of AI brought forth a new engineering science, "knowledge engineering," as an application-oriented one. Knowledge engineering (KE) is defined as the applied science of knowledge dealing, such as knowledge representation, knowledge utilization and knowledge acquisition. It aims the establishment of a knowledge information processing system, KIPS, with the aids of computers. As the basis of the approach the symbolic inference technique has been preferentially developed through the investigation of AI and cognition science, which handles recognition, decision and memory.

Knowledge representation is perhaps the most important area of KE research. It is the cornerstone on which all the other disciplines are built. Knowledge pertains to objects, relationships, and procedures in some domain of interest. Through the knowledge engineering study the following model have been presented: 1) rule-based model, 2) black board model, 3) causal network model, 4) frame-based model, and 5) semantic network model.

2.2. Expert Systems

The development of expert systems is one of the most exciting developments in the history of the computer; research, researchers, and products in this area are growing almost exponentially. Edward Feigenbaum had defined an expert system as: "... *an intelligent computer program that uses knowledge and inference procedures to solve problems that are difficult enough to require significant human expertise for their solution.*"

Though "expert system" is a more popular word in the application field, "knowledge system" should be acceptable as a standard scientific word. Knowledge systems are a class of computer programs that can advise, analyze, categorize, consult, and diagnose. They address problems that normally require the expertise of a human specialist. In a knowledge system, the rules (or *heuristics*) that are used for solving problems in a particular domain are stored in a *knowledge-*

base. Problems are then stated to the system in terms of certain facts that are known about a particular situation. The knowledge system then attempts to draw a conclusion from the facts using the knowledge–base. *Heuristics* are the rules of judgement that are used to make decisions from known facts. One of the major problems in the design of any expert/knowledge system is that of converting the knowledge and problem–solving techniques of the expert to a knowledge–base that can be used effectively to solve problems in the domain of expertise.

3. PHYSIOLOGICAL STATE CONTROL OF BIOREACTOR SYSTEMS

Since bioreactor systems inherently include complex biological networks, they are regarded as multi–variable and multi–structure systems from the view point of process control. There are many events which can qualify as structure–altering phenomena, such as changes of metabolic pathways, diauxic growth, changes in membrane transport mechanisms, morphological variations, expression or repression of genes in recombinant cells caused by chemical factors or temperature, spore formation, flocculation, etc. Since these physiological phenomena result in alterations of the internal structures of the systems, consequent alterations in the control system structure are required. We developed a multi–structure control system with a hierarchical structure, which is proposed for application to the control of bioreactor systems (Figure 2). It is capable of dynamic alteration of the control strategy according to the current process situation, generally the physiological situation of the cell population. The switching of the control strategy is done under the supervision of an additional module, which has the task of real–time analysis of the process in order to identify the physiological situation. The knowledge–based recognition system involves two distinct tasks. The first is the conversion of the input variables into features. The second is the translation of the feature membership to well–understood formal expressions.

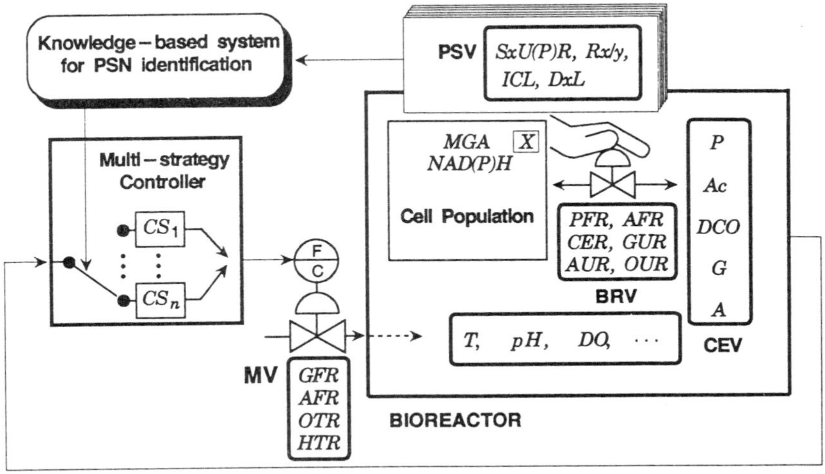

Figure 2. Structural scheme of a bioreactor system in a physiological control system.
MV: [*GFR*:glucose feed rate, *AFR*:ammonia feed rate, *OTR*:oxygen transfer rate, *HTR*:heat transfer rate].
CEV: [*G*:glucose, *A*:ammonia, *DO*:dissolved oxygen, *P*:end–product, *Ac*:acetic acid, *DCO*:dissolved carbon dioxide.
BRV: [*GUR*:glucose uptake rate, *AUR*:ammonia utilization rate, *OUR*:oxygen uptake rate, *PFR*:end–product formation rate, *AFR*:acetic acid formation rate, *CER*:carbon dioxide evolution rate].

4. CLASSIFICATION OF REPRESENTATIVE VARIABLES IN BIOREACTOR SYSTEMS

To elucidate the scheme for the physiological state control of bioreactor systems, looking first at the module for knowledge-based diagnosis to choose a control strategy, it is indispensable to clarify the structure of the bioreactor system and interpret the state variables involved in it. The first group of state variables in a bioreactor system, the variables of the input to the system, involve material and energy fluxes, some of which are represented quantitatively as flow rates such as the glucose feed rate (*GFR*), ammonia feed rate (*AFR*), acid/base feed rates, oxygen transfer rate (*OTR*), and heat transfer rate (*HTR*). These variables are called here *manipulated variables* (MVs). In the bioreactor, these inputs are transformed into a set of variables called the *cell-environmental variables* (CEVs). Typical examples of CEVs are the concentration of substrates such as glucose (*G*), ammonia (*A*) and dissolved oxygen (*DO*), and the concentration of various substances resulting from cellular activity, including the end-product (*P*), and by-products such as acetic acid (*Ac*) and dissolved carbon dioxide (*DCO*). Other types of CEVs are physical or chemical factors such as temperature (*T*) and pH. The cellular activities of consumption and production are quantitatively represented by the *biological rate variables* (BRVs), including the glucose uptake rate (*GUR*), ammonia utilization rate (*AUR*), oxygen uptake rate (*OUR*), end-product formation rate (*PFR*), acetic acid formation rate (*AFR*) and carbon dioxide evolution rate (*CER*). Physiologically, they describe flows that either enter or leave the cell.

We newly defined several groups of *physiological state variables* (PSVs). The first one is the specific rate of consumption or production, which can be defined by the formula *SxU(P)R* = uptake (or production) rate of a substance *x* / cell concentration. This group of PSVs includes the specific glucose uptake rate(*SGUR*), specific product formation rate(*SPFR*), specific oxygen consumption rate(*SOUR*), specific growth rate(*SGR*) and specific carbon dioxide evolution rate(*SCER*). The second is the ratio *Rx/y*, which is a large group of variables that are represented by the formula *Rx/y* = uptake (or production) rate of *x* / uptake (or production) rate of a substrate or a product *x*. As examples of uptake/uptake ratios, the ratio of the O_2 uptake rate to the glucose uptake rate (*R*o/g) and the ratio of the ammonia uptake rate to the glucose uptake rate (*R*a/g) are of significant importance for representation of the physiological state. Possible utilization of the production/production ratios are demonstrated by considering the ratio of the cell growth rate to the CO_2 production rate (*R*x/c), or the ratio of the ethanol production rate to the CO_2 production rate (*R*e/c) in ethanol fermentation. Variables of this type are known as yields. Examples of production/uptake ratios are the ratios of the cell growth rate to the glucose uptake rate (*R*x/g: *Y*x/g), to the O_2 uptake rate (*R*x/o: yield of cell from O_2), and the yield of cell from ammonia (*R*x/a). The most popular one is the respiration quotient (*RQ*). Another group is degrees of limitation (*DxL*), which can be useful in cultivations conducted under limitation by a particular product. This can be represented by the formula *DxL* = uptake rate of *x* / maximal potential uptake rate of *x*. Another group of interesting physiological variables involves intracellular levels (*ICLs*) of some important materials such as NAD(P)H, and glucose or its catabolite. We have developed a novel method to assess the availability of glucose in the cells, defining a special variable called a marker of glucose accumulation or *MGA*. *MGA* represents the delay (in seconds) from the moment of glucose feed rate interruption to the moment of DO increase. By developing a sensor of dielectric measurement of the microbial culture, we proposed a new physiological variable, *Q*d/x, specific capacitance (pF–L/g–cell), which may be utilized to monitor the change of physiological states of various different type of cells; bacterial, yeast, fungal, plant and animal cells.

5. PHYSIOLOGICAL SATE DIAGNOSIS BY A FUZZY INFERENCE APPROACH

The accumulated knowledge of the phenylalanine production process(12–14) allowed application of the knowledge-based control methodology. The synthesis of the control system is logically divided into two parts: first, the design of the higher system level (diagnosis subsystem) and, second, the design of the lower system level (control subsystem) as shown in the left upper part of Figure 2. The higher level is composed of procedure for calculation of the structural variables of the fermentation plant which represent adequately the physiological state of the cell population, and procedure for real-time diagnosis of the physiological situation (PSN) of the cell population. As global rules in a fuzzy inference approach for detection of the chronological (stage-associated) plant alterations, *if–then* rules were formulated to describe the plant structures, corresponding to the natural stages of the phenylalanine production process(15). A typical example is

IF (*RQ* is LOW) and (*MGA* is VERY HIGH) and (*SGR* is HIGH),

THEN (the process structural state is SCS_1: "STAGE 1: BALANCED GROWTH"); $dc_1 = 1.0$.

where SCS_i and dc_i are the abbreviations of subspace of constant structure and degree of certainty, respectively.

The translation of the linguistic expressions into a "transparent" algorithm yields a set of numerical equations for the calculation of the dc values from the membership values of all the conditions under consideration. According to the identified SCS, which is equivalent to the specified PSN, an appropriate control strategy is chosen from a set of candidates.

The described techniques was applied to diagnose and control the phenylalanine production process. The result from the work of the higher level of the control system is shown in Figure 3. Since the first two transfers had abrupt character, the static form of the rules did not hamper the recognition, and they were detected very reliably. However, the last transfer, which was gradual, confused the system and its detection was somewhat delayed. It came from the principle limitation of the static facts to interpret reliably alterations in the plant with more complicated dynamics. For better diagnosis, the corresponding rule should be enriched by dynamic facts.

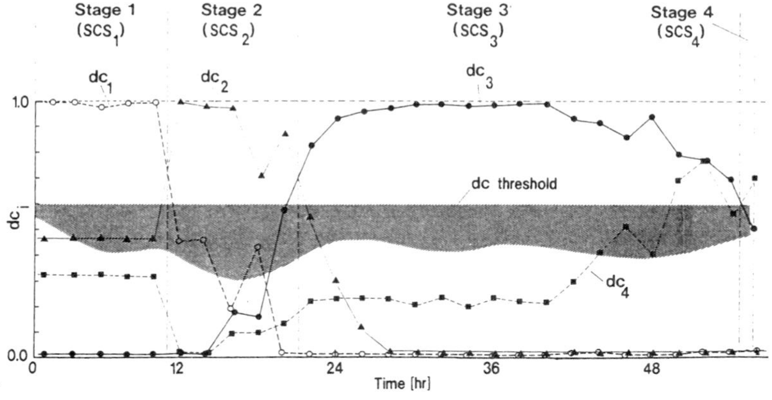

Figure 3. Results from the recognition procedure: time-courses of the degrees of certainty dc_1 – dc_4 used for detection of the stage transfers.

6. DETECTION AND CONTROL OF PHYSIOLOGICAL STATE VARIABILITY

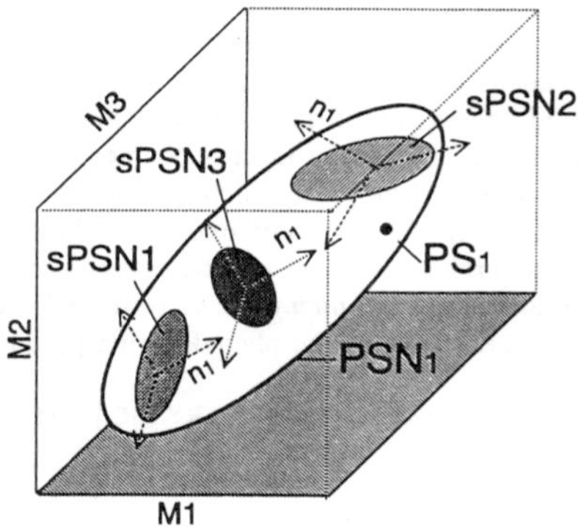

Figure 4. Mapping of physiological states (PS) in different domains or dimensions. Each physiological situation (PSN) or subsituations (sPSN) may be defined by a set of physiological variables (e.g. M1, M2, M3, n1).

In a structural space or physiological state space, there are several subspace of constant structure which are called physiological situation (PSN). Physiological situations are classified as primary regions of bigger scope in space, and these situations could represent the major characteristics and physiology of the organism (Figure 4). A transition in physiological state from one situation (e,g, PSN_i) to another situation (e.g. PSN_j) would imply a significant alteration in cell characteristics, which could also mean irreversible or partly reversible changes in the cell. For example, a recombinant organism loses its plasmid and results to a different characteristic in terms of nutrient consumption, growth pattern or product formation. In addition, within a given physiological situation, there can be one or more discrete subsituations which will be called physiological subsituations (sPSN). Physiological subsituations are parts of physiological situations and are similarly defined and composed by a set of physiological variables, but the type and meaning of these variables are quite different. The limited scope of physiological subsituations may indicate that these can defined by reduced number of variables, but this is not necessarily a prerequisite and depends on each subsituation. Physiological subsituations are suitable to describe minor, temporary or reversible changes in the physiological characteristics of the organism. The transition in physiological states among subsituations within a physiological situation would imply that the primary characteristics of the cell is basically retained but some properties are altered. For example, the physiological state or the cell may undergo transitions into subsituation corresponding to the production or excretion of metabolites. These state may be the response of the cells to environmental conditions or disturbances, while the fundamental characteristics are maintained. Elimination of the perturbations could result to another state transition to the former state, but it may not be an exactly identical physiological state.

6.1. Hierarchical Structure of Physiological States

The physiological state in a state space plane can be classified into a hierarchical structure based on specific criteria. This representation would be meaningful based on the process dynamics and state transitions that occur in a system of continuous culture of recombinants. The interrelationship among states can be clearly established such as the designation of causes and effects (Figure 5). The main physiological situations (PSN) are classified based on the ability of the cells to synthesize the product, which is related to the genetic and physiological characteristic of the organism. For recombinant organisms, the presence of plasmid containing the genes necessary for product synthesis is one prerequisite cell characteristic. The transitions in the main states are critical and viewed as irreversible or partly reversible, with respect to the total cell population, and depending on the cell characteristic. Plasmid–containing cells (X^+) that are altered and turned into plasmid–modified (X^*) or plasmid–deficient (X^-) non–producing cells is an example of irreversible transition. This transition in the cell population may be partly reversible if the plasmid–deficient cells do not proliferate or a control measure is implemented to prevent their favorable multiplication. Each of these state or situations are defined by a group of physiological variables. The other level consists of physiological subsituations (sPSN) which may occur within a physiological situation. Physiological events such as residual glucose accumulation (Subs+), acetic acid formation (HA+) or biomass washout (BM–) are considered reversible subsituations depending on their degree, since these events may appear or be suppressed by immediate control measures while the characteristics of the cell population remains in the same but not necessarily exact position in the physiological situation. Consequently, there is another group of physiological variables which describes these physiological subsituations.

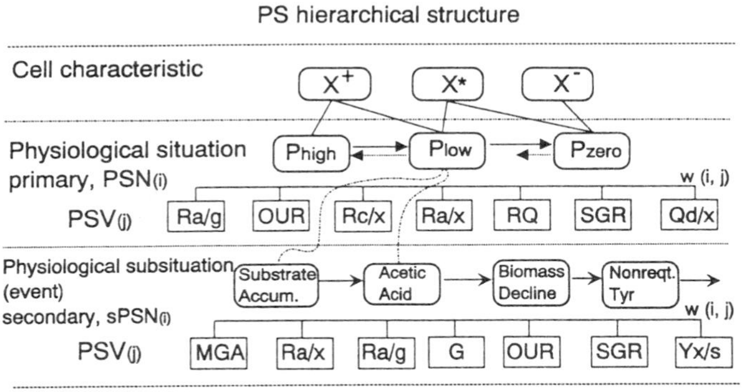

Figure 5. Schematic representation of the hierarchical organization of physiological situations (PSN) and its subsituations (sPSN) or events (PSE). Each state is described by a group of physiological variables (PSV).

6.2. Detection of Physiological States Transitions in Continuous Culture.

Identification of the physiological state of recombinant cell population in continuous culture was performed. The general physiological states and their transitions in continuous culture experiments are briefly discussed. The results of a series of experiments showed five main or primary physiological states were observed, namely; batch (B), transition state from batch to continuous (TS>C), stable state with phenylalanine production (PS–A), transition to another state (TS>S) and stable state without product formation (PS–B).

One factor related to the variability in physiological state was the alteration in the nutrient requirement of the cell, specifically the requirement for tyrosine in cell growth. This phenomenon was observed during continuous culture at a dilution rate, 0.1 h^{-1}, and glucose concentration of 50 g/L in the feed medium. After a short period of a transient state with stimulated phenylalanine production, a shift to another state with comparatively higher cell concentration and minimal phenylalanine production. The results of kanamycin–resistance test showed that the plasmid was retained in the entire cultivation period. The biomass level in the first physiological state (PS–A) showed that the amount of tyrosine (0.1 g/L) in the medium effectively restricted the cell concentration. In contrast with that of the following stable physiological state (PS–B), the cell level was higher than the theoretical yield from the amount of tyrosine that was added. This transition indicated the possibility that the characteristic of the host cell in terms of the auxotrophic requirement for tyrosine was altered, resulting to further utilization of the residual glucose present for growth.

The profile of some physiological variables showed distinct variation during the transitions in the physiological states (Figure 6). The specific growth rate (SGR) showed a peak during the transition state (TS>S) reflecting a steep increase in the cell concentration. The oxygen uptake rate (OUR) similarly showed a higher value in PS–B which was partly related to the cell concentration while the rate ratio of oxygen uptake to growth rate (Ro/x) decreased in the second physiological state (PS–B). The absence of product formation could be one factor that contributed to the decrease in Ro/x. The transition in the physiological state in this continuous culture exhibited a significant change in the characteristics of the cells. The identification procedure had clearly indicated these state transitions based on the high value of the certainty factor corresponding to the respective physiological states PS–A and PS–B, with a relatively high value, in the range between 0.4 and 0.7, of the certainty factor of the transient state, TS>S, for a short period before the start of physiological state PS–B.

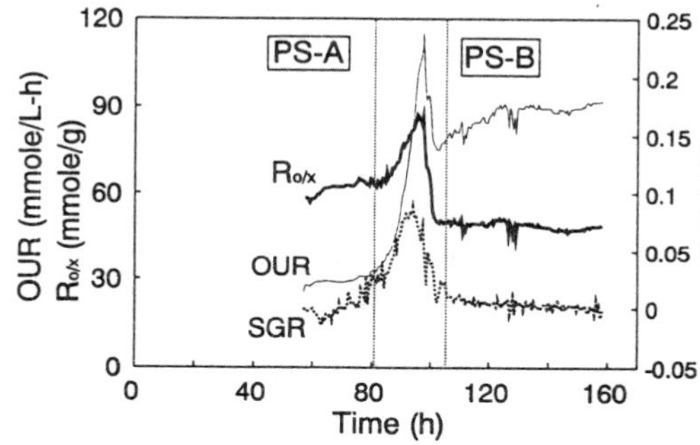

Figure 6. Profile of some physiological variables, namely; specific growth rate (SGR), oxygen uptake rate (OUR) and rate ratio of oxygen uptake to cell growth rate (Ro/x) during the transition un physiological state.

6.3. Control of Unfavorable State Transitions

The physiological state control approach was implemented in continuous culture of the recombinant *E. coli* by using a hierarchical control structure consisting of the higher level for detection of physiological state and supervisory control of the process using knowledge–base system. Control policies were designed and basically prioritized to attain the productive physiological state (PS–A), prevent the state transitions towards irreversible and unfavorable physiological situations, and minimize the presence of states within 'high risk' physiological situations which could exert temporary deleterious effect in the productive physiological state.

The objective during the physiological state at the transient period to continuous phase (TC>C) was to increase the cell concentration and attain the steady–state within a relatively short time period. The cell concentration level should be high but within the level that could still be provided with adequate oxygen supply by the reactor and could be maintained by the amount of tyrosine in the medium. The first control strategy (CS1) was implemented and the controlled variable was the glucose concentration in the feed medium (So) (Figure 7). The feedback control in the So of feed medium was dependent on the MGA value as input variable, and the MGA value was set at ca. 20–30 seconds in order to minimize over–limitation of glucose.

The stable productive physiological state (PS–A) was subsequently attained although there were slight variations in the cell concentration and occasional accumulation in acetic acid concentration (Figure 8). The peak of acid concentration at ca. 100 hours was the result of perturbation study to elevate the cell concentration by means of higher glucose feed concentration and investigate the cell transient response and product formation at higher glucose input rate. During the accumulation of acetic acid at this juncture, the important consideration was to determine the direction of the process and the priority of the control actions. The control objectives were to avoid the critical level of acetic acid concentration which could affect the growth of the organism, prevent the incursion into unfavorable physiological state and drive the cell population safely to the productive physiological state. The MGA control of the glucose feed rate could not adequately reduce the acetic acid level and the control policy was changed to the balanced DO–stat control strategy (CS2). In this control method the substrate feed rate (SFR) was the manipulated variable. This parameter was maintained at the set point by regulating the medium feed rate at fixed glucose feed concentration, and also by combined adjustment of both the parameters glucose feed concentration and dilution rate. Subsequently after ca. 110 hours cultivation, the control input was the shifted to the dilution rate and a constant glucose feed concentration was adopted. The detection procedure showed that the physiological state with acetic acid production decreased and gave lower a certainty factor. In addition, the cell physiological state was guided back to the productive state but not exactly identical with the

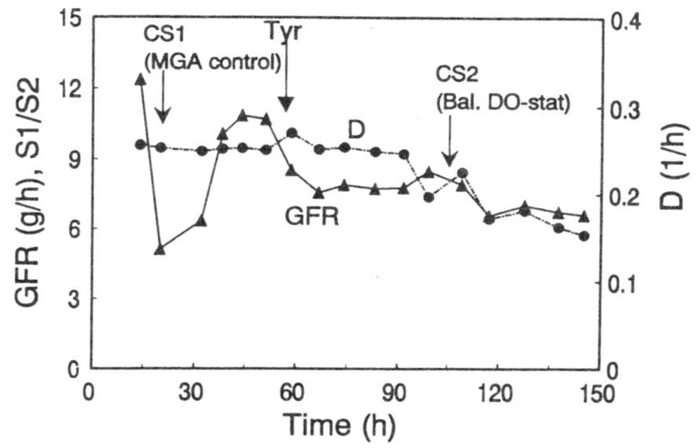

Figure 7. Profiles of controlled variables, glucose substrate feed rate (SFR, triangle) and dilution rate (D, circle). The arrows indicated the start of control strategies while the change in tyrosine concentration was indicated by Tyr.

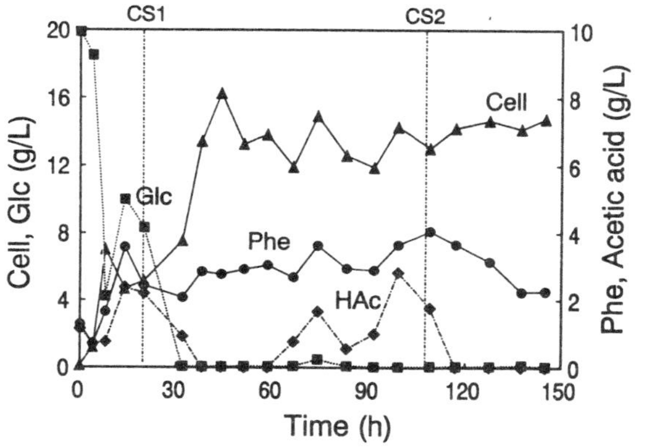

Figure 8. Time profiles of cell (triangle), phenylalanine (circle), acetic acid (diamond) and glucose (square) during continuous cultivation.

previous state.

7. CONCLUSION

The sensors available in a well instrumented bioreactor allowed on-line calculation of the following variables: *SGUR*, *OUR*, *SGR*, *R*o/g, *R*c/g, *R*x/g, *R*a/g, *R*o/x, *R*c/x, *R*a/x, *Q*d–x etc. We have found these variables informative, and have used them successfully in the development of a knowledge-based control system based on indirect physiological state control architecture (Figure 2). We have also introduced a method for detection and control of physiological state variability of biotechnological processes, and they have been successfully applied on the fed-batch cultivation and the continuous culture for phenylalanine production by a recombinant *E. coli*. Indeed, we believe that application of qualitative process theory to fermentation processes modeling (development of state-transition qualitative models) and control is very promising.

8. REFERENCES

1. Halme, A., 1989 Proc. IFAC Workshop AI in Real–Time Control, Shenyang, PRC 97–100.
2. Lubbert A., 1989 Proc. IFAC Workshop on Expert Systems in Biotechnology, Helsinki, December 11.
3. Aarts, R., Suviranta, A., Rauman–Aalto, P., Linko, P., 1990 Food Biotechnol. 4 301–315.
4. Iserman, R., 1987 Proc. 3–th IFAC Congress, Munich, FRG. 6.
5. Rod, M.G., Suski, G.J., 1988 Proc. IFAC Workshop on AI in Real–time Control, Clyne Castle.
6. Rod, M.G., Li, H., Su, S., 1988 Proc. IFAC Workshop on AI in Real–Time Control, Shenyang, PRC.
7. Cooney, C., O'Corner, G., Sanchez–Riera, F., 1988 Proc. 8th Int. Biotechnol. Symp., Paris 563–575.
8. Chen, Q., Wang, S., Wang, J., 1988 Proc. 4th Int. Symp. Computer Applications in Fermentation Technology, Cambridge, UK 253–261.
9. Linko, P., 1988 Ann. New York Acad. Sci. 542 83–101.
10. Konstantinov, K.B. and T. Yoshida, 1990 J. Ferment. Bioeng. 70 48–57.
11. Halme, A. and N. Karim, 1991 In K. Shugerl (Ed) Biotechnology 4: Measuring, Modeling and Control, Weinheim 625–636.
12. Konstantinov, K.B. and T. Yoshida, 1990 J. Ferment. Bioeng. 70 420–426.
13. Konstantinov, K.B., N. Nishio and T. Yoshida, 1990 J. Ferment. Bioeng. 70 253–260.
14. Konstantinov, K.B., N. Nishio, T. Seki and T. Yoshida, 1991 J. Ferment. Bioeng. 71 350–355.
15. Konstantinov, K.B. and T. Yoshida, 1992 J. Biotechnology, 24 33–51.
16. Perkins, W.A., and A. Austin, IEEE Expert 5 23–31.
17. Konstantinov, K.B. and T. Yoshida, 1991 IEEE Trans. Syst., Man, Cybern. SMC–21 908–914.

INDEX

C

F

G

H

I

J

K

L

N

O

P

R